SpringerBriefs in Electrical and Computer Engineering

For further volumes:
http://www.springer.com/series/10059

Roald Otnes · Alfred Asterjadhi
Paolo Casari · Michael Goetz
Thor Husøy · Ivor Nissen
Knut Rimstad · Paul van Walree
Michele Zorzi

Underwater Acoustic Networking Techniques

Roald Otnes
Maritime Systems Division
Norwegian Defence Research
Establishment (FFI)
PO box 115
3191 Horten, Norway
e-mail: roald.otnes@ffi.no

Alfred Asterjadhi
Department of Information Engineering
University of Padova
Via Gradenigo, 6/B
35131 Padova
Italy
e-mail: aasterja@dei.unipd.it

Paolo Casari
Department of Information Engineering
University of Padova
Via Gradenigo, 6/B
35131 Padova
Italy
e-mail: casarip@dei.unipd.it

Michael Goetz
Communication, Information Processing,
and Ergonomics (FKIE)
Fraunhofer Institute
(In cooperation with WTD71-FWG)
Neuenahrer Straße 20
53343 Wachtberg-Werthhoven
Germany
e-mail: michael.goetz@fkie.fraunhofer.de

Thor Husøy
Kongsberg Maritime
PO box 111
3191 Horten, Norway
e-mail: thor.husoy@konsgberg.com

Ivor Nissen
Research Department for Underwater
Acoustics and Marine Geophysics (FWG)
Bundeswehr Technical Centre for Ships
and Naval Weapons, Technology
and Research (WTD71)
Klausdorfer Weg 2-24
24148 Kiel
Germany
e-mail: IvorNissen@BWB.org

Knut Rimstad
Kongsberg Maritime
PO box 111
3191 Horten, Norway

Paul van Walree
Maritime Systems Division
Norwegian Defence Research
Establishment (FFI)
PO box 115
3191 Horten, Norway
e-mail: paul.vanwalree@ffi.no

Michele Zorzi
Department of Information Engineering
University of Padova
Via Gradenigo, 6/B
35131 Padova
Italy
e-mail: zorzi@dei.unipd.it

ISSN 2191-8112
ISBN 978-3-642-25223-5
DOI 10.1007/978-3-642-25224-2
Springer Heidelberg Dordrecht London New York

e-ISSN 2191-8120
e-ISBN 978-3-642-25224-2

Library of Congress Control Number: 2011942404

Printed on acid-free paper

Springer is part of Springer Science+Business Media (www.springer.com)

Preface

This SpringerBrief is a spin-off from the EDA (European Defence Agency) research project RACUN (Robust Acoustic Communications in Underwater Networks), which started in August 2010. RACUN has partners from the five countries Germany, Italy, Netherlands, Norway, and Sweden. The overall goal is to develop and demonstrate the capability to establish an underwater ad hoc robust acoustic network for multiple purposes with moving and stationary nodes.

One of the first research tasks in RACUN was a literature survey of state-of-the-art in underwater acoustic communication networks. When this work was done, it was decided that it would be a pity to keep a thorough literature survey on this rapidly emerging topic internal to the project. Therefore, we are glad to publish a slightly edited version of the RACUN literature survey as a SpringerBrief.

This literature survey presents an overview of underwater acoustic networking. It provides a background and describes the state of the art of various networking facets that are relevant for underwater applications. This report serves both as an introduction to the subject and as a summary of existing protocols, providing support and inspiration for the development of underwater network architectures. In recent years, other overview and survey papers have been published on the subject [1–6]. These papers can be consulted in addition to the present survey, which is however more comprehensive. Developments in the field of underwater sensor and communication networks are rapid, and new papers and protocols appear continuously.

The focus of this report is OSI layer 2 "Data Link Layer" and OSI layer 3 "Network layer". Several definitions can be found on the term "Link layer". In the OSI model, layer 2 "Data link layer" is split into two sublayers, MAC (medium access control) and LLC (logical link control). LLC is the upper of these sublayers.

After an introduction in Chap. 1, topics bordering the physical layer (time synchronization, full-duplex links, and adaptive data rate) are discussed in Chap. 2. MAC is discussed in Chap. 3, where considerations on frequency-division and code-division multiple access are followed by a detailed study on time-based multiple access technologies. Chapter 4 discusses logical link layer topics, including relatively new techniques such as fountain codes and network coding.

Chapter 5 gives an overview of routing (OSI "network layer"), including considerations on delay-tolerant networks.

The authors are affiliated with FKIE in Germany (Michael Goetz), WTD71-FWG in Germany (Ivor Nissen), University of Padova in Italy (Alfred Asterjadhi, Paolo Casari, and Michele Zorzi), Kongsberg Maritime in Norway (Thor Husøy and Knut Rimstad), and FFI in Norway (Roald Otnes and Paul van Walree). Due to the number of authors, it is inevitable that the writing style and level of detail is varying somewhat. Roald Otnes has been editing the report, and all the other authors are in alphabetical order in the author list.

Chapter 1 was written by Paul van Walree. Chapter 2 was written by Thor Husøy (Sects. 2.1–2.2) and Knut Rimstad (Sect. 2.3). Chapter 3 was written by Paul van Walree (Sects. 3.1–3.2), Michael Goetz (Sect. 3.3), Ivor Nissen (Sect. 3.3), and Roald Otnes (Sect. 3.4). Chapter 4 was written by Roald Otnes and Alfred Asterjadhi (Sect. 4.4.5). Chapter 5 was written by Paolo Casari, Alfred Asterjadhi, and Michele Zorzi.

In addition to the authors, the following helped in reviewing the original RACUN report: Jeroen Bergmans, Henry Dol, and Zijian Tang (TNO, Netherlands), and Svein Haavik and Jan Erik Voldhaug (FFI, Norway).

The RACUN project is part of the EDA UMS programme (European Unmanned Maritime Systems for MCM and other naval applications), and is funded by the Ministries of Defence of the five participating nations Germany, Italy, Netherlands, Norway, and Sweden.

References

1. Akyildiz IF, Pompili D, Melodia T (2005) Underwater acoustic sensor networks: research challenges. Ad Hoc Netw 3:257–279
2. Partan J, Kurose J, Levine BN (2007) A survey of practical issues in underwater networks. SIGMOBILE Mob Comput Commun Rev 11(4):23–33
3. Nguyen HT, Shin SY, Park SH (2007) State-of-the-art in MAC protocols for underwater acoustic sensor networks. Lecture Notes in Computer Science 4809/2007:482–493
4. Pompili D and Akyildiz IF (2009) Overview of networking protocols for underwater wireless communications. IEEE Commun Mag, 97–102
5. Shah GA (2009) A survey on medium access control in underwater acoustic sensor networks. In: Proceedings international conference on advance information networking and application workshops, WAINA'09, Bradford, UK, pp 1178–1183
6. Xiao Y (ed) (2010) Underwater acoustic sensor networks. CRC Press

Contents

Abbreviations

ACK	Acknowledgment
ACM	Adaptive coding and modulation
ACME	Acoustic communication network for monitoring of underwater environments in coastal areas
ADC	Analog to digital converter
ALBA-R	Adaptive load-balancing algorithm, rainbow version
ALOHA	Not an abbreviation, but a protocol name that means "Hello" in Hawaiian
ALOHA-ACK	ALOHA with acknowledgments
ALOHA-CS	ALOHA with carrier sense
AODV	Ad hoc on-demand distance vector routing
ARQ	Automatic repeat request
ASW	Anti-submarine warfare
ATM	Asynchronous transfer mode
aUT-Lohi	Aggressive unsynchronized tone-lohi
AUV	Autonomous underwater vehicle
BEB	Binary exponential backoff
BPSK	Binary phase shift keying
CDMA	Code division multiple access
CRC	Cyclic redundancy check
CSMA	Carrier sense multiple access
CTS	Clear to send
cUT-Lohi	Conservative unsynchronized tone-lohi
DAC	Digital to analog converter
DACAP	Distance-aware collision avoidance protocol
DAMA	Demand assigned multiple access
DBR	Depth-based routing
DBTMA	Dual busy tone multiple access
DS-CDMA	Direct sequence code division multiple Access
DSDV	Destination-sequenced distance-vector routing
DSR	Dynamic source routing

D-TDMA	Dynamic time division multiple access
DTN	Delay- and disruption-tolerant network
EPA	Expected packet advance
FAMA	Floor acquisition multiple access
FBR	Focused beam routing
FC	Fountain codes
FDD	Frequency division duplex
FDMA	Frequency division multiple access
FEC	Forward error correction
FFI	Norwegian defence research establishment
FIFO	First in first out
FPGA	Field programmable gate array
FSK	Frequency shift keying
FWG	Forschungsanstalt der Bundeswehr für Wasserschall und Geophysik
GF	Galois field
HDL	High-rate data link
HH-VBF	Hop-by-hop vector-based forwarding
IDMA	Interleave division multiple access
IRTF	Internet research task force
ISDN	Integrated services digital networks
KM	Kongsberg maritime
LDL	Low-rate data link
LLC	Logical link control
L-MAC	Low latency MAC
LT	Luby transform
MAC	Medium access control
MACA	Multiple access collision avoidance
MACA-U	Multiple access collision avoidance for underwater
MACAW	Multiple access collision avoidance wireless
MAI	Medium access interference
MANET	Mobile ad hoc network
MCM	Mine countermeasures
MFSK	Multiple frequency shift keying
MILD	Multiplicative increase linearly decrease
NAK	Negative acknowledgment
NCO	Numerically controlled oscillator
NTP	Network time protocol
OFDMA	Orthogonal frequency division multiple access
OLSR	Optimized link state routing
OSI	Open systems interconnection
PAGER	Partial-partition avoiding geographic routing
PCAP	Propagation-delay-tolerant collision avoidance protocol
PDR	Packet delivery ratio
PHY	Physical layer

PROPHET	Probabilistic routing protocol using history of encounters and transitivity
PSAM	Pilot symbol assisted modulation
QAM	Quadrature amplitude modulation
QPSK	Quaternary phase shift keying
RAPID	Resource allocation protocol for intentional DTN
Raptor	Rapid tornado
RF	Radio frequency
RTS	Request to send
SDMA	Space division multiple access
SDR	Software-defined radio
SEA swarm	Sensor equipped aquatic swarm
S-MAC	Sensor MAC
SNR	Signal-to-noise ratio
SRQ	Selective repeat request (= Selective Repeat ARQ)
ST-Lohi	Synchronized tone-Lohi
TDD	Time division duplex
TDM	Topology dicovery message
TDMA	Time division multiple access
T-Lohi	Tone-lohi
TNO	Netherlands organisation for applied scientific research
TP	Technical proposal
TSHL	Time synchronization for high latency
UT-Lohi	Unsynchronized tone-Lohi
UW	Underwater
VBF	Vector-based forwarding

Chapter 1
Introduction

1.1 Underwater Communications

Digital underwater communications are becoming increasingly important, with numerous applications emerging in environmental monitoring, exploration of the oceans, and military missions. Until the mid-nineties, the research was focused on hardware and on communication transmitters and receivers for the transmission of raw bits. In network terminology, this is known as the physical layer. A breakthrough was achieved in the mid-nineties by Stojanovic et al. [1], who showed that phase-coherent communication is feasible by integrating a phase-locked loop into a decision-feedback equalizer [2]. Such a receiver can be applied to a single hydrophone, although robust operation at high data rates, say >1 kbit/s, generally requires the presence of a (vertical) hydrophone array for reception. Indeed, multichannel adaptive equalizers have proven to be versatile and powerful tools. If the use of a receive array is impractical, as in multinode networks, then frequency-shift keying (FSK) is often used as a fairly robust modulation for single-receiver systems [3–5]. However, the corresponding data rates are of the order of 100 bit/s. Although progress is still reported on the physical layer, for example on multicarrier modulations or covert communications, a basic set of modulations and receiver algorithms is now available to support research on higher levels in network architectures.

Reasons for the rapidly increasing efforts put into research on underwater networks are various. The ongoing exploration of the oceans calls for sensor networks to support wide-area environmental monitoring. Military applications are also emerging [3, 6, 7], especially in the areas of autonomous sensor networks for mine countermeasures (MCM) and anti-submarine warfare (ASW). Autonomous underwater vehicles (AUVs) play an important role in such networks as they may replace traditional platforms for tasks in mine hunting or detection, classification, localization and tracking of a target. The advantages of AUVs include covertness, cost effectiveness and a reduced risk for personnel. Academia involved in the

R. Otnes et al., *Underwater Acoustic Networking Techniques*,
SpringerBriefs in Electrical and Computer Engineering,
DOI: 10.1007/978-3-642-25224-2_1,

development of radio frequency (RF) sensor networks are now discovering the underwater world, and the corresponding challenges and opportunities to publish original work.

1.2 The Acoustic Channel

The underwater acoustic channel is quite possibly nature's most unforgiving wireless communication medium [8]. Absorption at high frequencies, and ship noise at low frequencies, limit the usable bandwidth to between a few hundreds of hertz and tens of kilohertz, depending on the range. Horizontal underwater channels are prone to multipath propagation due to refraction, reflection and scattering. The sound speed of 1.5 km/s is low compared with the speed of light and leads to channel delay spreads of tens or hundreds of milliseconds. In certain environments, reverberation can be heard ringing for seconds and ultimately limits the performance of communication systems. The low speed of sound is also at the origin of significant Doppler effects, which can be subdivided in (time-varying) frequency shifts and instantaneous frequency spreading due to various mechanisms [9]. Both phenomena contribute to the Doppler variance of received communication signals, but require different measures at the receiver. A channel displaying both time-delay and frequency dispersion is known as a doubly spread channel. If the product of delay spread and Doppler spread exceeds unity, the channel is known as being overspread, and there is little hope for reliable communication at useful data rates.

Apart from covert applications, the performance of underwater communication systems is often not limited by noise but by multipath and Doppler, in the sense that it is not possible to increase the feasible data rate to an arbitrarily high value, or to lower the bit error ratio to an arbitrarily low value, by increasing the SNR. Acoustic modems used in underwater networks often operate at a data rate of no more than a few hundred bits per second. Higher data rates are certainly feasible, depending on the environment and conditions, but robust operation may then require a receiver equipped with an array of hydrophones.

There exist many different underwater acoustic communication channels, and this renders it difficult to design physical-layer solutions that are robust to area, weather conditions, and season. The following overview, which is not even complete, sketches the diversity of channels that can be encountered [9]. The channel may be characterized by correlated or uncorrelated scattering, by (quasi)stationary, cyclostationary, or nonstationary scattering. Shallow-water propagation channels range from stable monopath propagation (i.e., the ideal communication channel) to overspread, and from sparse to densely populated impulse responses. Doppler power spectra range from heavy-tailed to Gaussian, and may be symmetrical or skewed. The Doppler spread may be essentially the same for all paths, for instance for signaling through a sound channel, or vary by orders of magnitude in channels featuring a mixture of direct and surface-reflected paths. Signal fading can be Rayleigh, Rician, or K-distributed, while other channels may just as well be characterized as being deterministic [10]. In addition to signal propagation comes ambient noise, from many and varied sources. It can be colored and Gaussian when dominated by breaking waves, or impulsive and white when dominated by cavitating ship propellers or snapping shrimps. Further there are many noise sources featuring different statistics, such as precipitation, marine mammals, sonar systems, and offshore construction activities.

1.3 Networking

There is no single acoustic communication network that satisfies all needs and different applications require different approaches in many layers of the network. Network designers find themselves confronted with many challenges, as the acoustic medium is very different from the radio-frequency world above water. Owing to the dynamic environment of acoustic communication channels, the reliability of the physical layer is a major issue. There are many factors that can have a big impact on bit error ratio, package success rate, transmit power requirements, etc. For instance: passing ships, marine life, wind and waves, rain showers, seasonal cycles. Typical travel times between nodes are many orders of magnitude longer than in RF networks, and latency is one of the key factors in acoustic networks. Since the propagation delay is long, it is also highly variable when moving nodes are present in the network. When there are only stationary nodes the propagation delay varies by a few percent throughout the seasons. Energy efficiency is also an important design criterion [11] as the energy consumption of acoustic modems is generally in the range 1–100 watts while transmitting. The recovery of bottom nodes for battery recharge is a costly operation and modems are also a major burden for the limited battery capacity of AUVs. Another optimization criterion is network throughput. It is well-known that the data rates feasible with acoustic communications are lower than RF rates by orders of magnitude; the routing overhead of a network layer further reduces the net data rate. A factor of ten may be used as a rule of thumb for a multihop network topology.

So even though the basic principles of communication technology remain, underwater acoustic communications bring new and generally more challenging parameter regimes compared with RF communications. The result is that different (combinations of) techniques may be required in the two regimes.

Regardless of the functional demands and design priorities, medium access control (MAC) is an important ingredient of underwater networking. Without MAC, there is a high risk of collisions in a cacophony of unsolicited modem transmissions. Measures are needed to control the access of the medium by different users.

References

1. Stojanovic M, Catipovic J, Proakis JG (1993) Adaptive multichannel combining and equalization for underwater acoustic communications. J Acoust Soc Am 94(3):1621–1631
2. Stojanovic M, Catipovic JA, Proakis JG (1994) Phase-coherent digital communications for underwater acoustic channels. IEEE J Oceanic Eng 19(1):100–111
3. Freitag L, Grund M, von Alt C, Stokey R, Austin T (2005) A shallow water acoustic network for mine countermeasures operations with autonomous underwater vehicles. In: Proceedings Underwater Defense Technology (UDT), Amsterdam, Netherlands
4. Partan J, Kurose J, Levine BN (2007) A survey of practical issues in underwater networks. SIGMOBILE Mob Comput Commun Rev 11(4):23–33
5. Rice J, Green D (2008) Underwater acoustic communications and networks for the US Navy's Seaweb program. In: Proceedings 2nd international conference on sensor technologies and applications (SENSORCOMM), Cap Esterel, France, pp 715–722
6. Been R, Hughes DT, Vermeij A (2008) Heterogeneous underwater networks for ASW: technology and techniques. In: Proceedings Underwater Defense Technology (UDT), Glasgow, UK
7. Headrick R, Freitag L (2009) Growth of underwater communication technology in the US Navy. Commun Mag IEEE 47(1):80–82
8. Brady D, Preisig JC (1998) Wireless communications—signal processing perspectives. Prentice Hall, Upper Saddle River, Chapter 8, pp 330–379
9. van Walree P (2011) Channel sounding for acoustic communications: techniques and shallow-water examples. FFI-rapport 2011/00007, Forsvarets Forskningsinstitutt
10. Stojanovic M, Preisig JC (2009) Underwater acoustic communication channels: propagation models and statistical characterization. IEEE Commun Mag 1(47):84–89
11. Zorzi M, Casari P, Baldo N, Harris A (2008) Energy-efficient routing schemes for underwater acoustic networks. IEEE J Sel Areas Commun 26(9):1754–1766

Chapter 2
Topics Bordering the Physical Layer

2.1 Time Synchronization

The actual need for time synchronization within an underwater acoustic network is not always present. It can be argued that given a network with an operation time of hours or a few days, any standard equipment will have a clock drift that is negligible given most applications and network protocol stacks. Given this argument, synchronization of clocks can be done on board, before deployment. This might be true for some cases, even though from a practical and logistical point of view, especially when the number of nodes gets large, it gets time consuming to access all nodes individually through their electrical interface to set their clock manually. Another option is to synchronize clocks through switching the power on simultaneously for all nodes, but this can also be impractical. Common for both approaches is that the accuracy will vary and errors might occur (human in the loop). From this point of view it would be beneficial to be able to have the nodes doing time synchronization through the actual acoustic network.

The accuracy of a clock is inflicted by factors as temperature, supply voltage, shock [1] and ageing, all which an underwater network node is experiencing. The accuracy of the clock crystal is given in parts per million, ppm. Typical accuracies found in simulations for time synchronization methods are 40 ppm [1, 2], 50 ppm [3] and 80 ppm [4]. For example, given a clock with an accuracy of 40 ppm this effectively means a clock skew of 40 microseconds per second. Given an operation time of a week, the resulting clock offset will be 24 s. Such a figure might also result in a need for re-synchronization after deployment.

The application of the network is important when deciding the need for synchronization. Sensor networks might be divided into basically three categories in this respect [2]. The first group of applications merely requires the order of events, while the second requires the time interval of each of the events, whereas third require the absolute time of the event. Same type of division might also be true if any actuators are connected to the nodes. Delivery of packets in an underwater

R. Otnes et al., *Underwater Acoustic Networking Techniques*,
SpringerBriefs in Electrical and Computer Engineering,
DOI: 10.1007/978-3-642-25224-2_2,

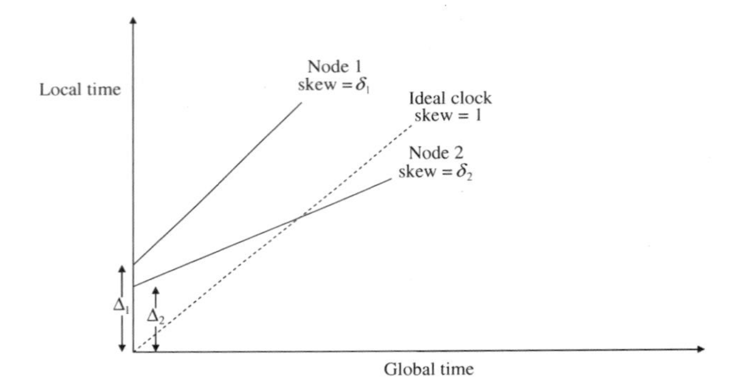

Fig. 2.1 Clock inaccuracies in two nodes

network generally has a high and non-deterministic latency, so time-stamping of sensor and actuator data with a global clock might be beneficial for many applications. Applications as target tracking and positioning requires time stamping. Also sleep scheduling for saving power will need time synchronization between nodes within a network. When it comes to the implementation of the network protocol stack, TDMA-based MAC protocol schemes benefit strongly from time synchronization.

2.1.1 Clock Inaccuracy Model

To avoid frequent re-synchronization between nodes it is beneficial to both estimate the clock skew and offset. The local time of any node i is related to the true global time, t by

$$t_i(t) = \delta_i \cdot t + \Delta_i$$

where, $t_i(t)$ denotes the local time of node i at time t, δ_i the clock skew and Δ_i the clock offset. Figure 2.1 illustrates the clock inaccuracies for two nodes. Generally it is assumed that clocks are short term stable, which is that they do not vary while doing estimation of clock skew [1]. This means that the clock drift can be represented with straight lines in the figure.

2.1.2 Time Synchronization Protocols

As for most of the other aspects in underwater networks, solutions and methods in wired and terrestrial networks can not be directly applied [1]. The Network Time Protocol (NTP) used for time synchronization on the internet copes with latencies but does no consider energy consumption issues. In RF-based sensor networks it is usually considered that the propagation delay is negligible, assuming nearly instantaneous and simultaneous reception and ignoring movement of nodes during synchronization. In underwater networks we know that propagation delays are large and variable.

Minimizing overhead of signalling for time synchronization is important due to the generally low data rate in underwater acoustic networks. The re-synchronization frequency should be minimized, thus the synchronization algorithm should be able to maintain a certain accuracy without the need for frequent re-synchronization. When re-synchronization is required the system performance should not degrade substantially. Any mobile nodes in the network introduce the need for the synchronization algorithm to compensate for the movements during synchronization.

Cross layer design with time stamping at the MAC layer is suggested by the work performed on synchronization within underwater acoustic networks [1, 2]. Utilizing data from the PHY layer in the protocol also shows to be beneficial [4].

Generally, what seems not to have been studied in detail in the literature found is how the synchronization is achieved network wide. Great amount of detail can be found on synchronization between two nodes or within a cluster, but things can get complicated when nodes start to move, extra nodes are deployed, or nodes are taken out of the network. Reference [2] is mentioning that a cluster needs to select its cluster head, but does not discuss it in detail. This might introduce some additional overhead for time synchronization, and is a point of further study. The degradation of time accuracy as a function of number of hops in a multi-hop network is suggested in [1] to degrade as the square-root of the number of hops. This is based on the assumption that the error per hop follows a Gaussian distribution of equal standard deviation. This might be a viable first order assumption.

The Mobi-Sync [3] method is one of the latest time synchronization protocols and is getting some attention for synchronization within networks with mobile nodes. What is special with this method is that it assumes that nodes are spatially correlated. That means that when one node moves, the other nodes also move in a related pattern. Even though this is the case for e.g. free floating drifters in a sea current, this generally does not hold when having gliders and AUVs in the network. The method also requires a dense network with every node having contact with at least three or more super nodes, a super node having correct time, in order to perform well. For static networks there are more energy efficient methods.

2.1.2.1 Time Synchronization for High Latency (TSHL)

The protocol TSHL [1] was proposed to compensate for high latency in acoustic networks. The method estimates and compensates both for clock skew and offset. This work assumes static nodes, and performance is strongly degraded when nodes are moving. It is shown in [2] that the method performs even worse than no

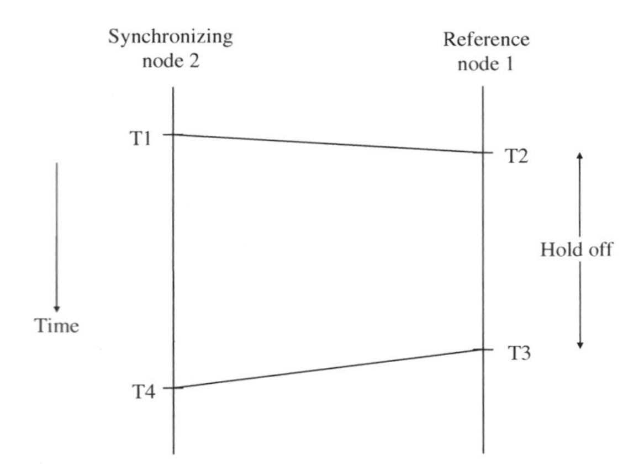

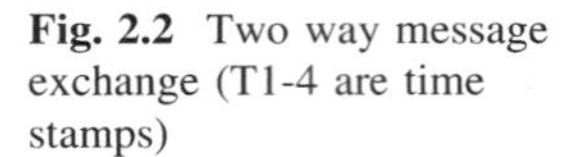
Fig. 2.2 Two way message exchange (T1-4 are time stamps)

synchronization when nodes move. This is due to the fact that the estimation of clock skew is inflicted by the movement.

This method is among the most energy efficient in the literature found. For estimation of clock skew, a beacon node is first broadcasting a number of messages to the neighbouring nodes. Every neighbouring node is then using linear regression on the times of arrival to estimate its clock skew. After that a single two way message exchange, see the scheme in Fig. 2.2, is used for estimating the clock offset. This is done as the reference node is informing the synchronizing node about time stamp T2 and T3 in the message sent back to the synchronizing node.

2.1.2.2 MU-Sync

The MU-Sync method [2] is designed for mobile networks. It assumes a cluster based network, and in contrast to the TSHL method it is the cluster head that takes responsibility for initiating and calculating the clock skew and offset for the nodes in the cluster. The cluster assumption does not exclude the method from working within a sparse network with maybe only one neighbour node to the cluster head.

The method is relying on two way message exchange for acquisition of clock skew and offset. The number of messages suggested is 25, same as for TSHL. The cluster head is then using linear regression to calculate clock skew and offset. Finally these parameters are distributed to each node.

2.1.2.3 D-Sync

Integrating the Doppler estimate of the PHY-layer for relative velocity estimates with the time stamps, preferably also at the PHY-layer, the D-Sync method [4] represents a novel approach for time synchronization in mobile underwater acoustic networks. Similar to Mu-sync it is the beacon or cluster head that initiates and calculates the clock skew and offset relying on two way message exchange. At the end the clock skew and offset is distributed to the synchronized node.

Reference is made to [4] for details of the method. There are two main sources of error in the method: the error due to Doppler measurements and the error due to the fact that Doppler measurements are not available continuously.

In a coherent transmission scheme accurate estimation of Doppler of the received signal is important to be able to equalize and decode the transmission. So the actual Doppler measurements tend to be very accurate. The work assumes a nominal error of 0.1 m/s, but simulations with up to 0.5 m/s are performed.

The Doppler is measured at T2 and T4 in Fig. 2.2. The time between these two measurements will in a dense network with a slotted contention based MAC protocol be governed by the Hold off time (T3–T2). This time might be several tens of seconds. This leads to a potential under-sampling of the Doppler and thereby the actual movement of the node. For slower moving nodes under water such as AUVs and gliders, this might not have such a severe effect. But it can be imagined that for gateway buoys on the surface submerged nodes in the splash zone exposed to wave motion, this under-sampling will degrade the performance of the algorithm.

Anyhow, simulations show that for a given set of parameters, including a network of 10 nodes distributed within a square of 1000 m sides, the error of the time sync two hours after the synchronization is 20 ms. This maps into an error of 2 s after a week of operation after the synchronization.

There is also described a light weight protocol B-D-Sync that has the same power consumption as TSHL. The performance of this protocol introduces a degradation of 5 times compared to the full D-sync.

2.1.3 Summary

Time synchronization is not always needed in an underwater acoustic network, but might be required given a long deployment, applications as target tracking or TDMA based protocols. Handling and logistics of nodes might also be simplified if they can be synchronized after deployment.

There exist a few time synchronization protocols in the literature. They all estimate both clock skew and offset in order to be able to minimize the need for re-synchronization. TSHL is suitable only for static networks, while Mu-Sync and D-sync are suitable for mobile networks. Even though designed for mobile networks it might be stated that even these methods would benefit from avoiding movement of nodes during synchronization.

All work on the methods considers local time synchronization between two nodes or within a cluster of nodes. Further work must be done to find optimal ways of getting network-wide synchronization in a multi-hop network.

2.2 Full-Duplex Links

A full-duplex link allows communication in both directions simultaneously. Full-duplex over the same physical medium is often emulated using the methods of Time-Division Duplex (TDD) or Frequency Division Duplex (FDD). TDD is bordering Time Division Multiple Access (TDMA) in functionality where separate time slots are used for sending and receiving signals. FDD is bordering Frequency Division Multiple Access (FDMA) where separate frequency bands are used for sending and receiving signals.

Full-duplex links are common in the cabled and radio frequency domain, included in systems as ADSL (cabled), UMTS (mobile) and satellite communication systems, while half-duplex links are predominant in underwater communication systems. No commercial full duplex modems seem to be available and a limited number of experiments has been conducted [5–7].

Obtaining full-duplex in a network is affecting the complexity of the link layer as well as the physical layer: The link layer may become simpler while the physical layer will be more complex.

2.2.1 Link Layer

Many half duplex underwater acoustic network protocols use collision avoidance by reserving the channel through a request to send and clear to send (RTS/CTS) session before accessing the channel. Further, flow control is often implemented using some kind of stop-and-wait flow control mechanisms. Given the large propagation delay of the acoustic channel this will lead to a lot of time waiting with potential low resource use efficiency. Study of channel reservation and flow control is done in Sects. 3.3.2 and 4.2, respectively.

If the available bandwidth is channelized and nodes are assigned unique channels within their respective two-hop neighborhoods, the need for collision avoiding coordination prior to message transmission is eliminated as each node is effectively operating over point-to-point links with its neighbors. The resulting full-duplex communications also allow for more efficient flow control mechanisms, such as sliding-window based methods [8]. The large propagation delay of the channel will result in the channel to act as a virtual buffer of data waiting to be read by the receiver.

The down-side of allocating two unidirectional channels for each connection is that it may result in very low bandwidth efficiency, unless the traffic from each node is regular and constant. However, in most data communications exchanges, the data is irregular and bursty, resulting in periods where allocated bandwidth is unused. This will again lead to potential low resource use efficiency.

In [8], this is mitigated by techniques from satellite communications capacity management: Demand Assigned Multiple Access (DAMA) controls and

Bandwidth-on-Demand (BoD) techniques. DAMA allocates channels to users when the users request an allocation. These channels are typically fixed in size. Alternately, BoD provides users a variable sized allocation depending on the request of the individual user. By allocating multiple channels to a user on demand, these channels may be used to inverse multiplex two or more message frames, thus providing a coarse version of BoD. It is this coarse BoD implemented via DAMA that is suggested.

2.2.2 *Physical Layer*

Time division duplexing is no option in an underwater network with the large propagation delays of the acoustic channel. Frequency division duplexing was demonstrated in [5] where data was transmitted between the shore and a ship. Using two transducers on the ship, one for transmission and one for reception, with a distance of 33 m between them, they managed to transmit and receive simultaneously in adjacent frequency bands. The distance between the ship and shore was up to 4500 m. Reception was good on both the ship and the shore. The frequency bands are not known, but Chebyshev filters with 80 dB out-of-band rejection were used for side-lobe suppression.

Obtaining full-duplex can also be achieved by using Code Division Multiple Access (CDMA) based techniques, see Sect. 3.2. In [7] a test using CDMA based channelization schemes was performed in a bucket and a small lake with a distance up to 5 meter. Separate Tx and Rx transducers were used with a spacing of 30 cm. Several channelization schemes were tested including frequency hop CDMA, time hop CDMA, direct sequence CDMA, and also hybrids thereof. Pulse position modulation was used for keying the data onto data symbols. In these tests, frequency hop CDMA performed best.

2.2.2.1 Transducers

All demonstrations utilize a separate transducer for transmitting and receiving signals, and apparently the most successful [5] having a long distance of 33 m between them. On a node like an AUV this kind of distance is not available, and preferably it should be a single transducer both transmitting and receiving. No literature could be found on full-duplex transducers. Radio systems like maritime VHF and maritime ship radio stations have FDD channels using only a single antenna.

2.2.3 *Concluding Notes*

Full-duplex has this far not been demonstrated to work well for underwater acoustic communications, and the required hardware is not commercially

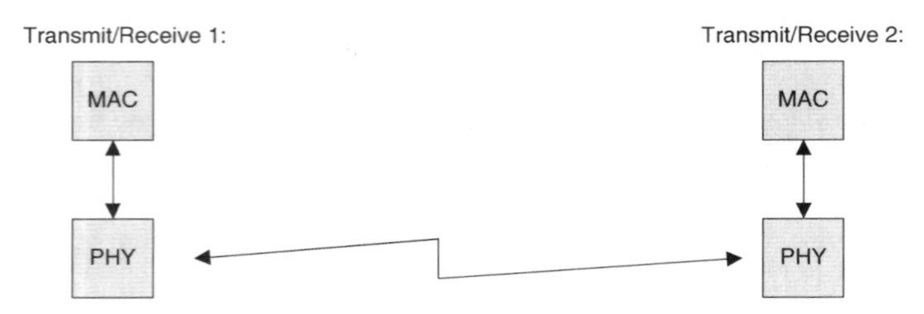

Fig. 2.3 The traditional MAC/PHY division (the OSI-model)

available. But if a good full-duplex solution is found in the future, it could significantly improve the performance of underwater acoustic network protocols.

2.3 Adaptive Data Rate

By Adaptive Data Rate in this context it is meant that the communication system is able to utilize some knowledge about the present state of the communication channel so that both the coding and modulation methods can be adapted to this state. The goal is to maximize the system throughput under varying channel conditions.

One way to achieve this is to employ a technique which in the telecom industry is known as "Adaptive Coding and Modulation", or ACM. This is used today in both wired and wireless communication systems.

In order to use adaptive coding and modulation effectively, it is necessary to establish a close interaction between the operations of the physical layer (PHY) and the medium access control layer (MAC).

In a traditional telecom environment, where circuit-switched networks (ISDN and ATM) were the norm, there were clear distinctive lines between the responsibilities of the PHY and the MAC, as it is laid out in the OSI model and as shown in Fig. 2.3.

With the introduction of packet switched networks (Ethernet and others), these lines have been considerably blurred, and this has lead to a simplified model where some of the layers have reduced functionality and others are removed.

In a communication scenario that involves ACM, this model will have to be changed further in that some of the traditional MAC functionality, such as the selection of the modulation format and coding scheme, will have to be moved down to the PHY layer in order to be able to respond to the (quickly) changing channel conditions. This part of the MAC functionality is sometimes referred to as the "lower-level MAC functionality" (Fig. 2.4).

In this scenario the "higher level MAC functionality" is responsible for establishing the overall system parameters like quality-of-service (QoS) requirements for the individual links, and to organize the network for maximum system capacity, the latter being important in an ad-hoc/mesh network scenario.

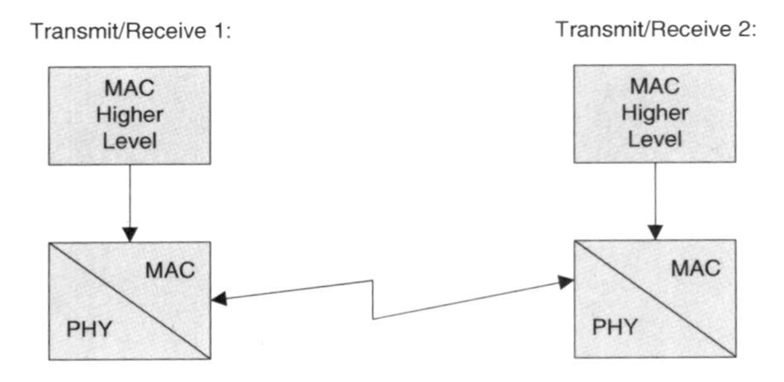

Fig. 2.4 A re-organized MAC/PHY division

In this discussion of adaptive data rate, only the PHY and lower-level MAC functionality will be considered, and it is easier to take the bottom-up approach and identify the requirements of the PHY first.

2.3.1 The Physical Layer

The discussion of the PHY-layer will be based on a conceptual transmitter/receiver pair, as shown in Fig. 2.5, below. Note that more detailed discussions on physical layer technology for underwater acoustic communications are not part of the present study.

In a communication system, each terminal will at least have one such transmitter/receiver pair (or "transceiver"). In addition, a transmit/receive switch circuitry is needed if time-division duplex (TDD) operation (not shown in the figure) is required.

In order to achieve maximum system capacity, this hypothetical system would have to be able to utilize all access techniques, such as frequency division multiple access (FDMA), time division multiple access (TDMA), code division multiple access (CDMA) and space division multiple access (SDMA). These access techniques are described elsewhere in this document, and will not be repeated here.

The various access techniques all have different requirements with respect to clock stability, frequency stability, linearity of up-/down-conversion chains and number and size of transducer elements, and will eventually be dictated by a cost/benefit trade-off.

A short description of the various functions carried out in the building blocks of Fig. 2.5 is given below.

From the information source comes the user data to be transmitted over the link. A forward error correction (FEC) encoder protects the data before transmission by adding special bits (parity) or bit-patterns that can later be utilized in the receiver to extract the original information bits. There are a number of different coding schemes that can be applied for this purpose, and they range from simple to advanced block-coding structures (from Hamming to Reed-Solomon) to convolutional and Turbo codes, and various concatenations of the above.

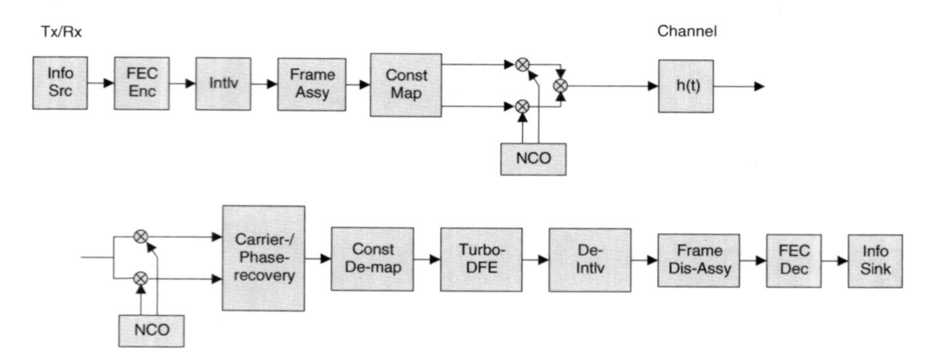

Fig. 2.5 Conceptual transmit/receive functionality. See text for explanation

In the interleaver, the encoded data are repositioned in the data-stream according to a predefined structure. This is to avoid loss of data caused by impulsive noise which would be detrimental to a convolutional code decoding process. (This is not so much of a problem with block-codes.)

In the frame-assembly block, the information is inserted into a frame where pre-ambles like unique words (for frame alignment) or various pilot-assisted modulation (PSAM) bits and post-ambles like CRC's and/or end-of-frame (EOF) delimiters are placed.

In the constellation mapper, the individual bits of the frame are mapped onto a suitable alphabet of symbols, later to be modulated onto two orthogonal wave-forms (for a quadrature modulated signal).

For a coherently modulated signal, the alphabets can be a set of symbols belonging to a modulation format like e.g. binary phase shift keying (BPSK), quaternary phase shift keying (QPSK) and higher order like quadrature amplitude modulation (QAM).

For a code division multiple access system based on a direct sequence spread spectrum (DSSS) technique the alphabet is a set of orthogonal spreading codes.

For a non-coherently modulated signal, the alphabet is e.g. a set of frequencies in a multiple frequency shift keying (MFSK) modulation format.

A code division multiple access system can also be constructed by using a set of frequencies in a frequency hopping pattern orthogonal for each code.

The symbol set is then modulated onto two orthogonal carrier waveforms (cosine and sine), where the carrier frequency is generated through a numerical controlled oscillator (NCO). In Fig. 2.5, a direct-to-carrier type of system is shown. The transmitted signal, $s(t)$, is a real band-pass signal.

At the receive side, the signals are converted directly from carrier (real band-pass signal) to complex base-band, the demodulation frequency (and phase, in case of a coherent demodulation scheme) are again controlled by a NCO. The exact frequency and phase are controlled by the carrier-frequency- and phase-recovery sub-system.

Not shown in the figure are the necessary signal/burst acquisition sub-systems.

The information bearing signals are then extracted from the signal constellation and fed to an equalizer to remove inter-symbol-interference (ISI) induced by the channel.

In the frame disassembly operation, the information bearing signal is extracted and fed to the de-interleaver before FEC decoding and the result is then fed to the end-user.

In [9], it is claimed that a coherent modulation scheme based on Phase Shift Keying (BPSK/QPSK), in conjunction with an adaptive decision feedback equalizer (DFE) and a spatial diversity receiver is an effective way of combating the effect of multipath fading in a shallow water environment. It is, however, admitted in the article that the excessive delay spread, often several hundred symbols, make it too computational complex for real-time operation for such a system.

2.3.2 Medium Access Control, Lower Level

In order to be able to utilize the communication channel effectively, an optimally configured system will have to be able to change both the modulation format and the coding scheme in order to adapt to the current channel conditions.

To make this work, it would be necessary to collect information about the current channel conditions, convey this information back to the transmitter side and use this to select the modulation and coding schemes for the next outgoing burst.

The information to be used in this process could be e.g. an estimate of the signal-to-noise ratio and/or the delay and Doppler spread of the channel, and this information could be extracted from known symbols in the frame structure like the pre-amble (e.g. a unique-word) or the PSAM symbols (if they are used).

However, any addition of extra symbols to aid in these processes would incur an overhead and thus reduce the information data rate.

Other, more indirect means of extracting the same information could be to monitor the states of an equalizer or the decoding depth of the trellis in an (convolutional code decoder) FEC decoding process.

A different way of obtaining information about the channel state could be to use a special burst sent in advance as a channel probe, as proposed in [10]. In their system, such a probe is sent just before the actual information burst is transmitted, as shown in Fig. 2.6.

Here station 1 is transmitting a probe before the actual data is transmitted, and this is used at station 2 to extract some key properties of the channel state at the moment of reception. This information is then affixed to the data in the next outgoing burst from the station 2. If the turn-around time is short this is a more or less correct description of the channel, and it is used at station 1 for the next outgoing burst from this station.

Fig. 2.6 Using a channel probe to extract information about the channel

In addition, if the channel can be treated as reciprocal, station 2 can itself use this information to set the parameters for the next outgoing burst, and/or to construct a better probe for more detailed channel measurements.

The approach described in the paper is a very simple adaptation of this, but it lacks a conclusive statement about the system improvements, if any, that can be achieved using this technique. It is not clear why a special probe is required to do this, as the same result can be achieved by simply embedding the same information into the pre-amble of the burst itself.

2.3.3 Adaptive Data Rate in ARQ Systems

Adaptive Coding and Modulation is something that is normally closely connected to the physical layer processes of a communication system, as described in the previous sub-sections. This is due to the fact that very low latency is required in order for the system to respond to the changing channel conditions.

If ACM methods based on direct measurements of the current channel conditions are not possible, e.g. where the time-delays in the system are so severe that the channel measurements are obsolete by the time the system can respond to these, an ACM approach based on the use of information extracted from the ARQ system (see Chap. 4) can be foreseen.

One way to achieve this could be that instead of using an approach where the previous packet is blindly repeated when an ACK time-out occurs or a NAK is received, one would use a scheme where the first re-transmission is just a re-transmission of the packet. If this transmission is also unsuccessful, the packet is reformatted into a longer burst where the modulation and coding schemes are strengthened (more energy per bit and/or more protection bits added).

If the transmission still fails, this back-off procedure is repeated until all possibilities are exhausted and the system finally breaks down.

When the ACK packets start to arrive, the modulation and coding overhead is gradually reduced until the maximum possible information bandwidth is re-established on the channel. The transmit side will always be aware of the current maximum channel capacity based on the reception of the ACK-packets, since this will confirm that the receiver is able to decode the packets.

2.3.4 Summary and Conclusions

In terrestrial wireless communication systems, ACM has been in widespread use for some time. But as is pointed out in [11], the key difference between a terrestrial radio-based communications system and an underwater acoustic communication system is the large propagation delay, low bandwidth and high bit-error-rate. Their conclusion is that a direct adaptation of a terrestrial radio protocol may not provide acceptable results, and that protocols for this purpose need to be developed from the ground up.

The research into adaptive data rates in underwater communication systems seems to be in its infancy, and in order to move this research forward and in the end create a successful system employing ACM the following factors needs to be addressed:

The special requirements for an underwater acoustic communication system need to be taken into account already at the design stage. There are currently no existing solutions that can directly be applied.

Measures for estimating the (strongly time varying) channel impulse response needs to be built into the burst structure in order to reduce measurement latency and overhead in the communication link.

In order to have maximum flexibility in the selection of possible signalling waveforms to use at any given time, a software defined radio (SDR) approach is necessary. In any case, lessons learnt in the SDR arena should be taken into consideration in a new design.

In order to be able to realize a fully software defined radio, the hardware platform need to be as flexible as possible, and that means that all the necessary functions must be realized in digital domain. Especially important in this respect is to ensure that the digital-to-analog (DAC) and analog-to-digital (ADC) conversion subsystems have the speed and precision needed.

Fortunately, the latter is steadily improving with the advances in ADCs, DACs, field-programmable-gate-arrays (FPGA) and microcontrollers (uC), and in the end the transducers and the required signal conditioning circuitry will be the limiting factors.

References

1. Syed AA, Heidemann J (2006) Time synchronization for high latency acoustic networks. In: Proceedings of 25th IEEE international conference on computer communications INFOCOM, Barcelona, Spain, pp 1–12
2. Chirdchoo N, Soh W-S, Chua KC (2008) Mu-sync: a time synchronization protocol for underwater mobile networks. In: Proceedings of ACM WUWNet, San Francisco, CA, USA
3. Liu J, Zhou Z, Peng Z, Cui JH (2010) Mobi-sync: efficient time synchronization for mobile underwater sensor networks. Technical report UbiNet-TR10-01, UCONN CSE
4. Lu F, Mirza D, Schurgers C (2010) D-sync: doppler-based time synchronization for mobile underwater sensor networks. In: Proceedings of ACM WUWNet, Woods Hole, MA, USA

5. Jarvis S, Janiesch R, Fitzpatrick K, Morrissey R (1997) Results from recent sea trials of the underwater digital acoustic telemetry system. In: Proceedings of 7th international conference on electronic engineering in oceanography, Southampton, UK, Halifax, Nova Scotia, pp 186–192
6. Smith KB, Larazza A, Kayali B (2001) Scale model analysis of full-duplex communications in an underwater acoustic channel. In: Proceedings of MTS/IEEE Oceans, Honolulu, HI, USA, (4):2250–2255
7. Xie GG, Gibson JH, Bektas K (2004) Evaluating the feasibility of establishing full-duplex underwater acoustic channels. In: Proceedings of 3rd annual Meditteranean ad hoc networking workshop (MedHoc), Bodrum, Turkey, pp 449–459
8. Gibson J, Xie G, Kaminski A (2005) Demand assigned channel allocation applied to full duplex underwater acoustic networking. J PACON International, pp 81–95
9. Stojanovic M, Catipovic J, Proakis JG (1993) Adaptive multichannel combining and equalization for underwater acoustic communications. J Acoust Soc Am 94(3):1621–1631
10. Benson A, Proakis J, Stojanovic M (2000) Towards robust adaptive acoustic communications. In: Proceedings of MTS/IEEE Oceans, Providence, RI, USA, (2):1243–1248
11. Chitre M, Shahabudeen S, Stojanovic M (2008) Underwater acoustic communications and networking: recent advances and future challenges. Marine Technol Soc J 42(1):103–116

Chapter 3
Medium Access Control

Medium access control (MAC), also known as multiple access control, is a sub-layer of the data link layer and manages access to the medium. In underwater networks, MAC protocols orchestrate the access to the acoustic communication channel. Without MAC, collisions of unsolicited modem signals may greatly degrade the overall network performance. The basic MAC objective is to avoid collisions, but more generally MAC protocols deal with network throughput, latency, energy efficiency, scalability, and adaptability. Weights can be given to different MAC objectives, depending on application and requirements. MAC protocols can be subdivided in contention-free schemes and contention-based schemes.

Contention-free schemes are directly linked to the physical layer and avoid collisions by assigning different frequency bands, time slots, or codes to different users. The nodes in such a network do not actively compete with one another in order to obtain access to the medium. The three basic types are illustrated in Fig. 3.1 for a scenario with three active nodes: frequency-division multiple access (FDMA), time-division multiple access (TDMA), and code-division multiple access (CDMA). The theoretical net data flow is the same for these cases, but the physical-layer consequences, feasibility, and real-world performances may be very different. A related technique, known as space-division multiple access (SDMA), uses phased arrays at the transmitter and/or receiver side to allow spatial separation of nodes. SDMA is outside the scope of this report.

Contention-based MAC protocols avoid pre-allocation of resources to individual users, i.e., nodes in the network. Instead, the users compete with one another to obtain medium access on demand. Minimization of collisions is the key task of the MAC layer, while at the same time keeping the required overhead within bounds. The remainder of this chapter presents an overview of existing MAC protocols, starting with the contention-free schemes.

R. Otnes et al., *Underwater Acoustic Networking Techniques*,
SpringerBriefs in Electrical and Computer Engineering,
DOI: 10.1007/978-3-642-25224-2_3,

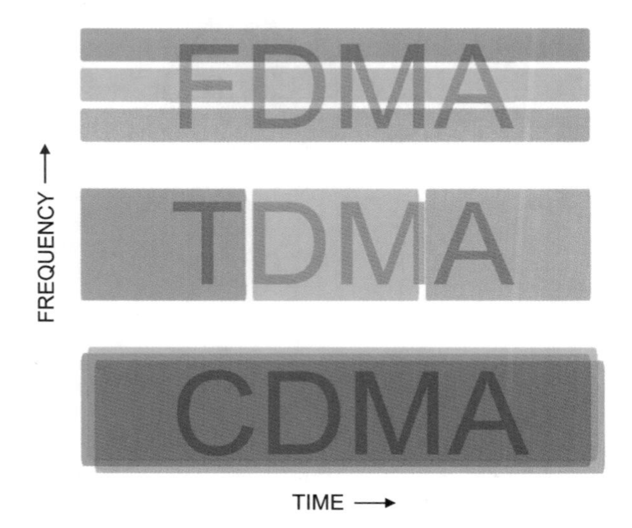

Fig. 3.1 Conceptual scheme of contention-free multiple access schemes

3.1 Frequency-Division Multiple Access

3.1.1 Description

Frequency-division multiple access (FDMA) is a potentially contention-free medium access scheme that assigns different frequency bands or discrete tones to different users. This allows nodes to transmit and receive at the same time, in the same overall frequency band, without interfering with one another. FDMA faces difficulties in underwater networks, owing to the physics of acoustic communication channels. The available bandwidth is limited, and multipath propagation causes deep spectral nulls. A network with many nodes implies a narrow frequency band for each node, and a high risk of complete fading of some nodes in the network. For this reason, FDMA is considered as being unsuitable for underwater acoustic networks [1, 2]. Another argument against FDMA is the difficulty to realize narrowband filters and to suppress sidelobes of users in neighboring frequency bands. This is further complicated by the near-far problem described in Sect. 3.2.2. However, when discrete tones of different users are orthogonal to one another, such as in OFDMA, this is not necessarily a problem. Furthermore, FDMA should be kept in mind as a solution for "hybrid" networks consisting of clusters with short intracluster node distances and long intercluster distances. Within a cluster, a high frequency band can be used, whereas the longer ranges between clusters are served with a lower frequency regime. Different types of modems (transducers) will be required for such a system.

Note that FDMA is not necessarily contention-free: If each node is assigned a transmit frequency, several nodes may still attempt to address the same destination simultaneously. Then there is contention unless the destination is able to receive at several frequencies simultaneously. If each node is assigned a receive frequency, there is contention when more than one node tries to access the same destination.

3.1.2 Case Studies

A simple case of FDMA was used in a sea experiment in 1997 to achieve full-duplex communication between the shore and a ship [3]. Using two transducers on the ship, one for transmission and one for reception, with a distance of 33 m between them, they managed to transmit and receive simultaneously in adjacent frequency bands. Reception was good on both the ship and the shore.

Early phases of Seaweb utilized FDMA [4]. In Seaweb'98 for example, 5 kHz of acoustic bandwidth was subdivided into 120 discrete MFSK bins. The modulation supported three interleaved sets of 40 MFSK tonals, but in order to reduce MAI (medium-access interference), half the available bandwidth was unused. Thus, only 20 tonals were available per user. The users, in this case, were three clusters. Within a cluster, TDMA was used, while FDMA was the MAC choice for intercluster communications. Seaweb'99 made a step forward in the direction of network self-configuration by permitting the server to assign FDMA receiver frequencies. The paper [4] mentions an inefficient use of bandwidth as a drawback of FDMA. The net bit rate was 50 bit/s during the Seaweb'99 experiments. FDMA was employed in Seawebs'98 an'99 primarily for ease of implementation. Seaweb 2000 started pursuing hybrid CDMA/TDMA methods.

3.2 Code-Division Multiple Access

3.2.1 Description

Many sources consider code-division multiple access (CDMA) to be a promising physical and MAC layer technique for underwater networks. The basic principle is the use of (binary) codes to modulate the information stream in a spread-spectrum fashion. When different nodes use different codes with low cross-correlation, in the same overall frequency band, their data can be received simultaneously by other nodes in the network. The main advantage over FDMA is robustness with respect to frequency selective fading, since the entire frequency band can be used by all nodes. The main advantage over TDMA is that the medium can be accessed simultaneously by all nodes. The price paid is that low cross-correlations generally require long codes, and long codes imply reduced data rates. Moreover, correlation properties of codes are degraded by Doppler effects in the acoustic channel. CDMA delegates part of the network design problem back to the physical layer, which requires complex algorithms for multi-user detection and demodulation.

Direct-sequence CDMA and frequency-hopping CDMA are well-known candidate modulations. Recently, a scheme has been considered whereby different users are distinguished by different interleavers in (turbo) coded systems [5]. This method is called interleave-division multiple access (IDMA), but can also be seen as a subcategory of CDMA.

3.2.2 Near-Far Problem

CDMA is most effective when the sound pressure levels are of comparable magnitude at the receiver. If one node is much weaker than another, the interference due to the strong node may hamper detection or demodulation of the weak node. This is known as the near-far problem, as nearby nodes are normally received stronger than faraway nodes. Solutions are 1) to let the network dynamically adjust transmit power levels or 2) the use of very long codes (and thus low data rates).

The designation "near-far problem" is slightly misleading, because the distance is only one of several factors that influence the sound pressure level at the receiver. This is illustrated by the following data from the ACME project [6]. Experiments were performed in the Bay of Douarnenez in 2002. This bay was a quiet environment with a relatively flat bathymetry at a $\sim$20 m water depth. Three slave modems (R1, R2, R3) were placed on the seafloor as shown in Fig. 3.2, from [7]. The network was configured in a TDMA fashion [8], allowing the master modem to receive sensor data from each node in different time slots. At the position of the master modem there was also a vertical receive array with three hydrophones. The received time series show that the "strength" of a node cannot be predicted on basis of the distance alone. The relative strength of nodes depends strongly on the receiver depth. If CDMA was to be used in this environment, the network would face the task of adjusting source levels and/or code lengths to ensure manageable sound pressure levels at reception. If the receiver would be an AUV with a variable range and/or depth, the network would constantly be busy updating these settings. This is comparable to the problem of TDMA with moving nodes, where the network constantly needs to update time tables.

3.2.3 Case Studies

An early CDMA-based MAC protocol was sketched by Xie and Gibson [9]. They describe a centralized scheme in which the master node generates a tree topology and updates routes, codes, power levels, etc., at fixed time intervals. This limits RTS/CTS exchange and reduces the propagation delay along each route. The scheme allows new nodes to be added dynamically, and allows dead nodes to be removed from the structure. The scheme controls the transmission range of each

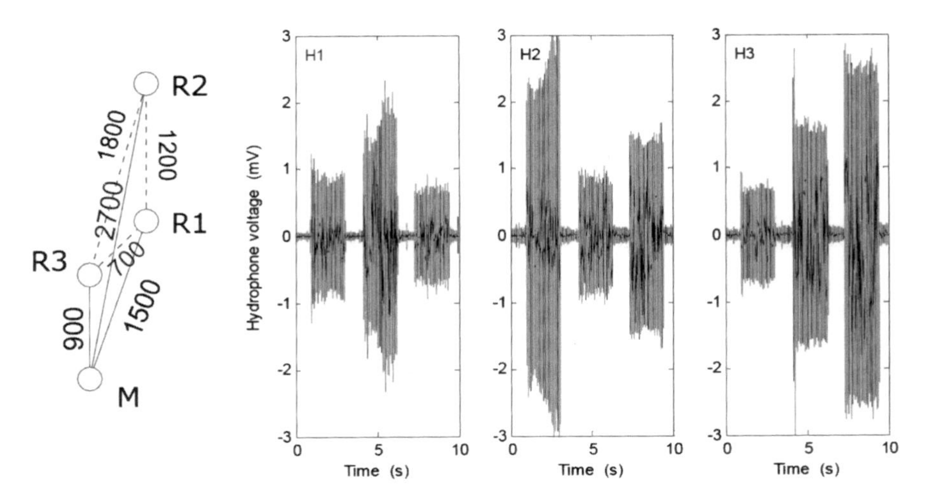

Fig. 3.2 *Left* Network topology with three bottom nodes (R1, R2, R3) and a master modem (M) with a vertical receive array. The numbers are distances in meters. *Right* TDMA reception on three hydrophones (H1, H2, H3) of signals transmitted by, respectively, R3, R1, R2

node and permits reuse of code words in different geographical parts of the network.

Tan and Seah [10] propose a distributed CDMA-based MAC protocol for underwater sensor networks. It involves a three-way handshake (RTS-CTS-DATA), but collects RTSs from multiple nodes before issuing a single CTS that addresses all these nodes. Subsequently, the nodes respond with data and their signals can be received simultaneously owing to the CDMA properties. Simulations show that the proposed protocol offers higher throughput than Reservation ALOHA and MACA (see Sect. 3.3.2).

Pompili et al. recently proposed a CDMA-based MAC protocol for underwater networks [11]. The protocol, UW-MAC, accomplishes multiple access to the scarce acoustic bandwidth and aims at achieving three objectives: 1) high network throughput; 2) low channel access delay; 3) low energy consumption. In deep water (little multipath), it simultaneously meets these objectives. In multipath environments, it dynamically finds the optimal trade-off between the objectives according to the application requirements. UW-MAC combines CDMA with ALOHA, incorporates an algorithm to jointly set the optimal transmit power and code length, and does not rely on handshaking algorithms such as RTS/CTS. Simulations show that UW-MAC outperforms competing MAC protocols under all considered network architecture scenarios and settings.

Another recent CDMA MAC protocol [12], also named UW-MAC, aims at achieving low latency for applications such as intruder detection. The protocol uses a combination of DS-CDMA and TDMA on the MAC layer, clustering algorithms, and spatial reuse of codes. It features a sleep/wake-up schedule that minimizes power wasted on idle listening and reduces the end-to-end delay of messages. It even takes into account chemical properties of batteries to enhance node and network lifetimes. Simulations show a reduced latency and enhanced throughput of UW-MAC compared with S-MAC (sensor MAC), L-MAC (low latency MAC), and MACA.

3.3 Time Based Multiple Access Technologies

In *time based multiple access* a network node uses for a transmission the complete bandwidth for a certain time period. As a result, the signals are more resistant to frequency-selective fading than in frequency based multiple access strategies like FDMA [13]. Multiple transmissions have to be done one after another to avoid collisions due to overlapping signals at the receiver, like shown in Fig. 3.3.

A transmission T is characterized by a start time t_S and an end time t_E where:

$$t_S, t_E \in \mathbb{R} \text{ with } t_S < t_E$$

and transmission length τ is given as:

$$\tau = t_E - t_S.$$

Collisions occur, if a transmission t_1 from node A overlaps with a transmission t_2 from node B at the receiver node C:

$$\left[t_{1_S} + p_{A,C}, t_{1_E} + p_{A,C}\right] \cap \left[t_{2_S} + p_{B,C}, t_{2_E} + p_{B,C}\right] \neq \emptyset$$

where $p_{x,y}$ is the signal propagation delay between node x and y. If the *signal-to-interference-and-noise ratio* (SINR) of the overlapping signal is too high, the signal can not be decoded and the packet is lost.

3.3.1 Study of Existing Strategies

In this section we introduce the different strategies to avoid such collisions and give an overview about the existing *medium access control* (MAC) protocols for *time based multiple access*. They can be divided into the following categories:

A *Time Division Multiple Access* A time interval, a so called frame, is divided into time slots of fixed length ω. Each time slot is assigned to a network node wherein the node is allowed to send. The complete transmission must be finished within this time slot:

$$t_S, t_E \in [s, s + \omega] \text{ with } s = s_0 + l \cdot n, n \in \mathbb{N}$$

where s_0 is the start time of the first assigned time slot and l the frame length (Fig. 3.4).

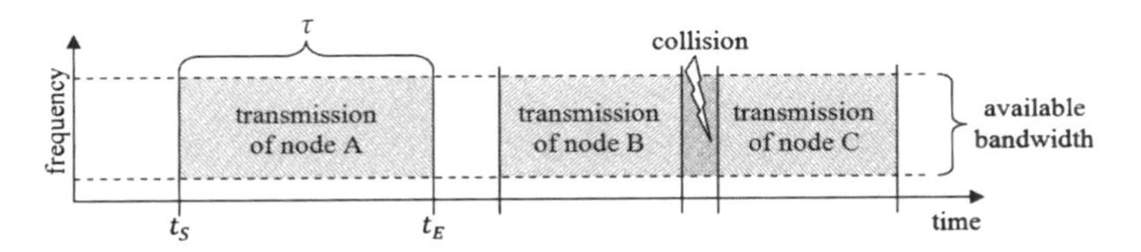

Fig. 3.3 Scheme of time based multiple access

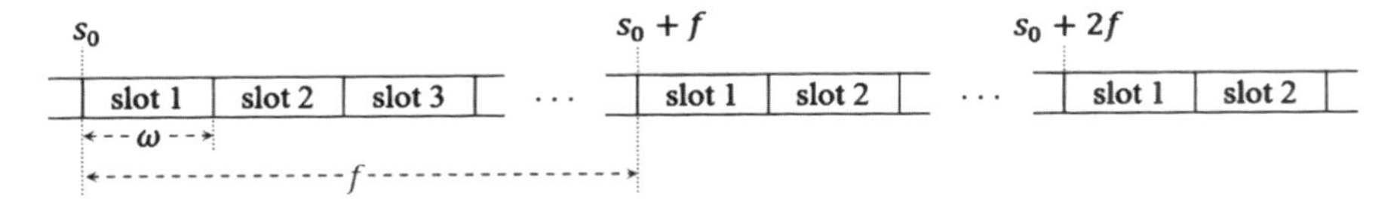

Fig. 3.4 Scheme of time division multiple access

B *Random Access* In this strategy the selection of the transmission start time t_S and end time t_E is arbitrary:

$$t_S, t_E \in \mathbb{R} \text{ with } t_S < t_E.$$

Random access protocols may be subdivided into:

B.1 *Direct Random Access*: Send data packets directly without preceding channel reservation.

B.2 *Channel Reservation:* Reserve the channel with short control packets or control tones, before transmitting data packets.

C *Slotted Access:* All transmissions are synchronized to reduce the partial overlapping of colliding packets. For this purpose the time is divided into slots of length ω and a transmission start time t_s is deferred to the beginning of the next time slot:

$$t_S = t_0 + \omega \cdot n, n \in \mathbb{N}$$

where t_O is a synchronized generic start time of the first slot.
To avoid overlaps between slots the transmission time is limited to τ_{max} with:

$$\tau_{max} = \omega - p_{max}.$$

where p_{max} is the maximum propagation delay.
Note that a slot is not assigned to a single user like in case A and collisions occur if multiple nodes select the same time slot. Instead the slotting improves the maximum achievable channel utilization of random access protocols from category B [14, 15].

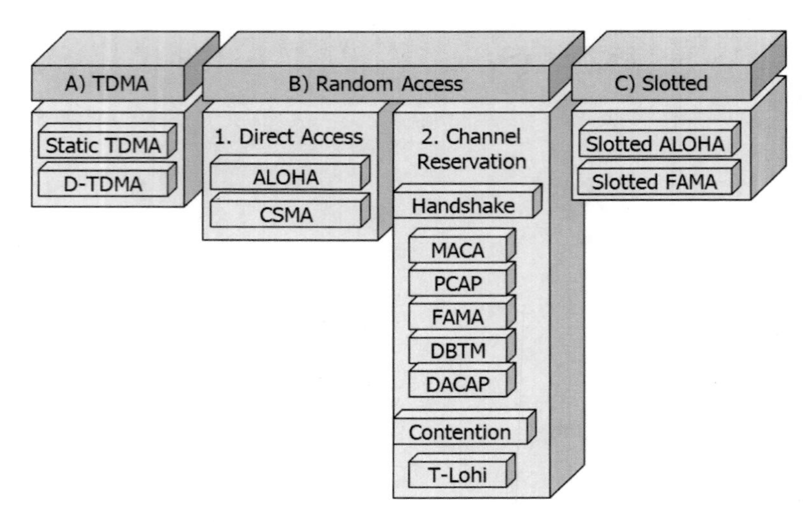

Fig. 3.5 Family tree of time based multiple access technologies

3.3.2 *Study of Existing Technologies*

In the following sections we describe the different *medium access control* (MAC) protocols structured by the introduced categories (Fig. 3.5), evaluated to the aspects of capability for underwater acoustic networks.

A. Time Division Multiple Access

In *Time Division Multiple Access* (TDMA) a time interval, a so called frame, is divided into time slots of fixed length ω which are repeated cyclically, like shown in Fig. 3.4. The slot timing is generic in the network; hence all nodes must be synchronized in time. This can be done in advance or underwater during an initialization phase (see Chap. 2.1), for example via the Mobi-Sync algorithm proposed by Liu et al. [16], though this needs additional energy and time.

In TDMA each time slot is assigned to a single network node that can use the entire frequency band during this time slot. To guarantee a contention free communication, guard times are included between each time slot, dependent on the maximum propagation delay and synchronization accuracy [17], as shown in Fig. 3.6. The next time slot must not start before the data packet of the previous slot is propagated to all neighbour nodes to eliminate the possibility of partial overlaps.

As a consequence, the full bandwidth cannot be used and the channel utilization is given as:

$$Utilization = \frac{\omega}{\omega + p_{\max} + \Delta}$$

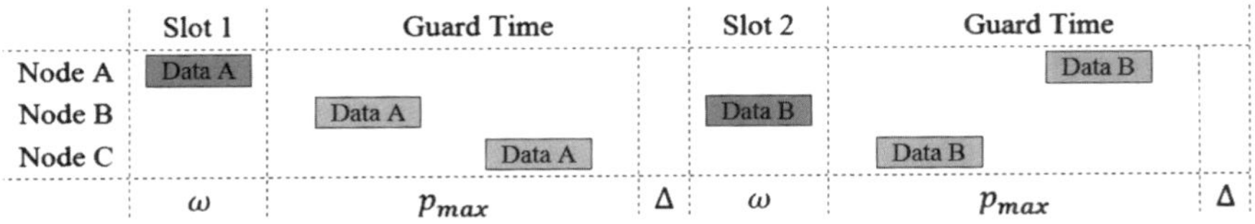

	Slot 1	Guard Time		Slot 2	Guard Time	
Node A	Data A				Data B	
Node B		Data A		Data B		
Node C		Data A			Data B	
	ω	p_{max}	Δ	ω	p_{max}	Δ

Fig. 3.6 TDMA channel access scheme with guard times

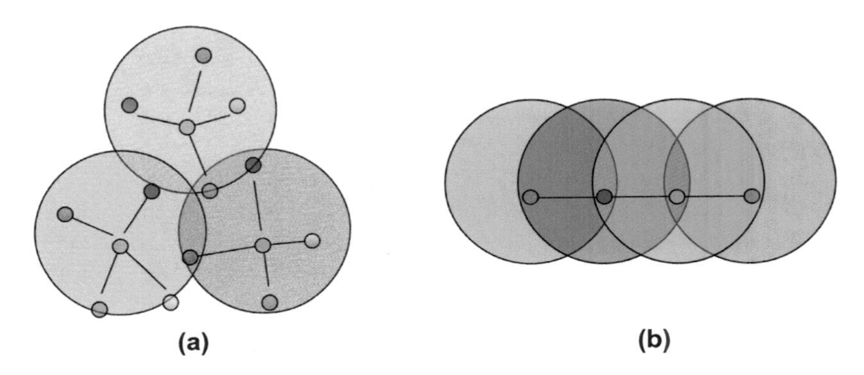

Fig. 3.7 Static TDMA with different frequency bands for each cell. **a** Centralized slot allocation **b** Distributed slot allocation with spatial reuse

where p_{max} is the maximum propagation delay, Δ the maximum synchronization difference and ω the slot length during which nodes are allowed to transmit their data. Due to the slow sound propagation in underwater acoustic networks, TDMA is not suitable for long-range networks with high transmission ranges.

The slot scheduling in TDMA can be done centralized by a so called master node or distributed by each node itself. The first one is the most common version, where a centralized master node, for example a gateway buoy, administrates the slot allocation, like applied in the acoustic local area network (ALAN) deployed in Monterey Canyon, California [18] or in the mine counter measure networks with AUVs presented in [19]. A master node and its assigned slave nodes build a so called cell, where each slave must be in the transmission range of the master node. To enable multi hopping, single cells can be combined to a network cluster, where the master nodes are typically connected via a backbone network, for example gateway buoys can use radio links for this. To reduce the intercell interferences between neighbouring cells, TDMA can be combined with FDMA, where each TDMA cell gets its own frequency band, for example used in GSM, shown in Fig. 3.7a. Time slots, visualized with the colour of the nodes, can be reused in each cell, because each neighbouring cell uses its own frequency band, marked as the background colour of the cell.

In the second version of the slot scheduling, each node must choose its time slot itself. Thereby, slots can be spatially reused, like in the centralized version, if the transmission ranges of nodes with the same slot do not overlap, otherwise collisions can occur at nodes in the overlap area. This is shown in Fig. 3.7b, where the third node (green) must use another slot than the first node (red) to guarantee a collision free communication at the intermediate node (blue). Only the fourth node can reuse the slot of the first node (red) without causing collisions. The slot allocation problem was shown in [20] to be equivalent to the *graph colouring problem*, where each neighbouring node must have a different colour, by adding to the graph additional edges between each 2-hop neighbour. The graph colouring problem is known to be NP-complete, also for planar graphs [21]. In [22], Chlamtac introduces an algorithm for the distributed slot allocation in dynamic multihop networks. The algorithm is highly localized, because each node only needs to know its 2-hop neighbourhood and thereby approximates a good slot distribution.

By the way of allocating slots in TDMA, there are not only differences in who makes the slot allocation, but also when the allocation is done. Each node can be assigned a slot once at the time it enters the network or dynamically on demand, when a node has data ready to send. The two different variants of TDMA are presented in the following sections.

A.1 Static TDMA

In static TDMA the assigned time slots of each node are preconfigured or allocated when entering the network, from a master node or itself in the distributed versions.

Resulting of the fixed slot length, each node possesses a guaranteed data rate, given as:

$$User\ Data\ Rate = Utilization \cdot \frac{Total\ Data\ Rate}{\#slots}$$

The user data rate is equal for each node, but can be extended for individual nodes by assigning it multiple slots, for example to relay or master nodes with higher traffic load.

A guaranteed data rate is an advantage if nodes send periodically data packets or there is the need of a *quality of service* (QOS) to guarantee a data rate and a maximum channel access delay. The average channel access delay is given as:

$$Average\ Access\ Delay = \frac{1}{2} \cdot \#slots \cdot (\omega + p_{\max} + \Delta) = \frac{1}{2} \cdot Cycle\ Length$$

In consequence, the slot length ω is a trade-off between channel access delay and channel utilization. In context of underwater networks a high slot length must be chosen, proportional to the maximum propagation delay, to get acceptable channel utilization. Therefore TDMA is limited to networks with delay tolerant applications and low network utilization [23]. Furthermore, the number of slots is a trade-off between the channel access delay and the maximal number of supported slaves in a cell, respectively the maximal number of allowed 2-hop neighbours in the distributed case. Moreover, the channel utilization is lowered if there are more slots than nodes and some slots are left unallocated, which is inevitable if the node density is diverse. To optimize the channel utilization, the cycle length can be calculated adaptively, but cycle changes are associated with several problems [24]. Due to the fact that the cycle length must be generic in a cell respectively in the complete network, the nodes must agree on the new size and on the time in which the new cycle should go into effect simultaneously at all nodes. Methods for reaching such a consensus are discussed in [25]. Another approach for distributed networks was suggested in [26] by Kanzaki et al., where unallocated slots are dynamically assigned to nodes in this area. If new nodes enter, the proposed protocol generates unassigned slots by depriving one of the multiple slots assigned to a node. In the case that no more slots can be freed, the cycle length is doubled, but only in this local area. Therefore, nodes can have neighbours with different cycle lengths.

A.2 D-TDMA

If the network traffic is relatively static, for example voice data, the fixed time slotting of TDMA is an advantage to guarantee a fixed data rate and channel access delay. But if the network traffic is event-based and bursty, some nodes have nothing to send and slots are idling while other nodes want to transmit greater data bursts over multiple slots. In *Dynamic TDMA* (D-TDMA) [27, 28] nodes can request a variable number of slots for a designated time on demand, when they have data ready to send, transmitting special control packets during a defined control slot. All slots can be assigned in this way, or only a subset of additional slots besides fixed allocated slots, which guarantees a minimum data rate but allows also enhancing the data rate on demand (for example, if the network traffic consists of voice and event-based data packets).

B. Random Access

If the generated network load is not periodic and uniformly distributed over all nodes, but instead event based and bursty, generated from few nodes, it can be more efficient to grant each node the full bandwidth for a variable time. But if multiple network nodes are assigned to the same channel, signals from different nodes can overlap at a receiver and may result in packet losses. Furthermore, most underwater modems are half-duplex, therefore packets arriving during a transmission are lost too.

This section is an overview of the existing methods for random medium access in such shared channels, subdivided in protocols with and without preceding channel reservation as categorized in the introduction of this chapter.

B.1 Direct Random Access

Medium access protocols of this category send directly their data packets without a preceding handshake for channel reservation. An optional carrier sensing before transmitting avoids disturbing ongoing transmissions from other nodes. If the channel is currently busy, a transmission is deferred and repeated at a later point in time.

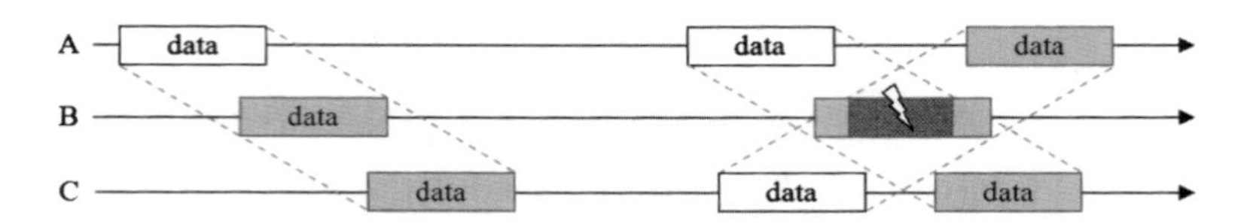

Fig. 3.8 ALOHA channel access scheme with packet collision

B.1.1 ALOHA

The simplest method to access the medium is sending immediately whenever a node has data to send, as used in the ALOHA protocol. In the original version [29], neither channel sensing nor retransmission was implemented. ALOHA does not use any collision avoidance strategy and packet loss can occur, as shown in Fig. 3.8.

In an adaptation called *Aloha with carrier sense* (ALOHA–CS), each node senses the carrier before sending its data and waits if necessary until the channel is free. But carrier sensing does not guarantee collision freeness, especially not in environments with high propagation delays, as shown in Fig. 3.8. Here, both node A and node C sense the channel as free before transmitting, but there is still a collision at node B.

In terrestrial environments where RF is used and the signal propagation delay is short, carrier sensing can lead to synchronization of nodes, because all waiting nodes transmit simultaneously as soon as the channel is free. Due to the significantly higher signal propagation delay in underwater networks, it is not clear if such effects also occur in underwater environments. The simulations in [30] showed that ALOHA-CS delivers results that are slightly worse than ALOHA in all configurations.

To detect collisions, ALOHA can be enhanced with acknowledgment packets (ALOHA-ACK), which are sent back by the receiver if the packet is received successfully. A node will not transmit the next packet until it receives the acknowledgment of the current packet. If packets are lost and no acknowledgement arrives during a designated period, a timer runs out and the sender waits a random backoff time to avoid new collisions and then retransmits the packet, as shown in Fig. 3.9.

In networks with high network utilization, ALOHA becomes inefficient causing many collisions, and therefore wastes energy. On the other side, ALOHA shows good performance in networks with low utilization or long transmission ranges. Furthermore ALOHA can be a good choice for impulse communications, where the packet transmission times are short and collisions unlikely.

B.1.2 CSMA

The *Carrier Sense Multiple Access* (CSMA) protocol uses carrier sensing like ALOHA-CS. For CSMA there exist different strategies to mitigate the synchronization effects, whereby multiple nodes, waiting for a free channel, access it simultaneously after the channel is detected as free, resulting in collisions. In the

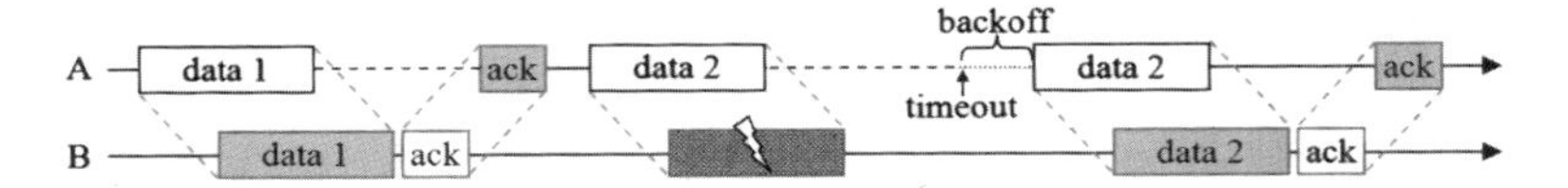

Fig. 3.9 ALOHA with acknowledgements scheme with retransmission

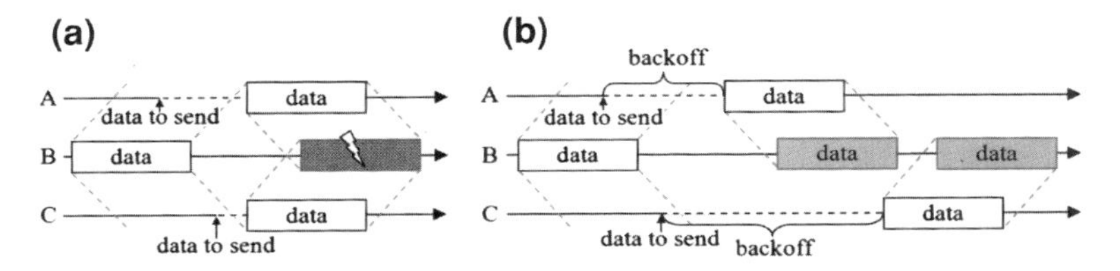

Fig. 3.10 Carrier sensing with and without random backoff. **a** ALOHA-CS (without backoff). **b** CSMA (with backoff)

first version, the so called *p-persistent CSMA*, with $0 < p \leq 1$, a node only sends with a probability of p when the channel is detected as free. With the probability of 1-p the node waits a defined backoff time, before sensing again. If the channel is still free it will again send with a probability of p. The special case *1-persistent CSMA* is equivalent to ALOHA-CS. The second version, also known as *non-persistent CSMA* additionally includes random backoffs, a multiple of the maximum propagation delay, to mitigate the probability of collisions due to synchronization during multiple nodes wait for a free channel, as shown in Fig. 3.10.

CSMA was developed for terrestrial radio networks and it has to be proven if the backoff algorithm also reduces collisions in underwater networks with high propagation delays. If nodes are not uniformly distributed, such that the propagation delays differ significantly compared to the transmission times, collisions due to synchronization effects may be rare. Nevertheless, some node constellations may be vulnerable to this effect in underwater networks too.

B.2 Channel Reservation

In this kind of approach, nodes at first reserve the channel before transmitting their original data packets. This can be done via control packets or special control tones in a subchannel. The preceding reservation shall reduce collisions of data packets and thereby compensate the additional traffic and delay through the control packets respectively tones. But this is only the case if the data packets are long and there is a lot of traffic in the channel and therefore the possibility of collisions of large packets is high.

In this section we describe different reservation strategies and their advantages and disadvantages in case of underwater communication. The protocols are subdivided into the following two categories:

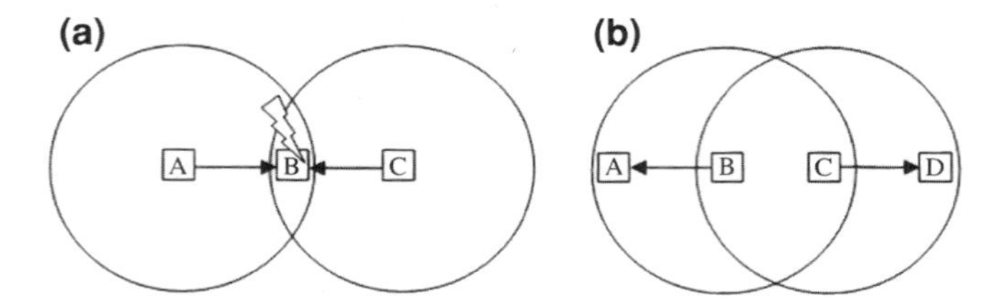

Fig. 3.11 **a** Hidden node problem and **b** exposed node problem

B.2.1 *Handshake-based reservation:* Reserve the channel with short control packets or tones during a so called handshake phase.

B.2.2 *Contention-based reservation:* In advance of each data transmission all nodes with data ready to send compete with each other for channel access during a so called contention round. The contention will be repeated until there is a winner.

B.2.1 Handshake-based Reservation

These protocols use a handshake to reserve the channel before sending data packets. A transmitter sends (instead of the data packet) a control packet or a tone to inform the receiver and all neighbours that it has data to send and wants to reserve the channel for its transmission. The receiver replies, if the channel is free and not reserved for other transmissions in the receiver's neighbourhood. After receiving that the channel is reserved, the transmitter can start the transmission of the data packet. This handshake shall guarantee freeness of strong interfering signals from neighbours, which would influence the SNR at the receiver, during the complete transmission of the data packet.

Since collisions occur at the receiver, carrier sensing at the transmitter does not guarantee collision freeness, called the *hidden* and the *exposed node problem* [31]. This can be resolved with handshakes. In the *hidden node problem* the transmission of one node is hidden to another node resulting in collision at the receiver as shown in Fig. 3.11a where the circles around each node shows the transmission ranges. If node A transmits data to node B, this transmission is hidden to node C, because it is out of A's transmission range. Therefore carrier sensing at node C does not guarantee collision freeness at node B. During a handshake node B informs its neighbourhood with the reservation reply that node A is going to send a data packet, so node C defers its transmission. In the *exposed node problem* a transmission is deferred though it is not necessary as shown in Fig. 3.11b. If node B transmits data to node A, all nodes in B's neighbourhood defer their transmission if they use carrier sensing. But if node C wants to send data to node D it is not necessary to wait, because node C does not interfere the reception at A and node B not the reception at D. Using handshake node C must only defer its transmission if it overhears the reservation reply from node A, otherwise it knows A is out of the transmission range.

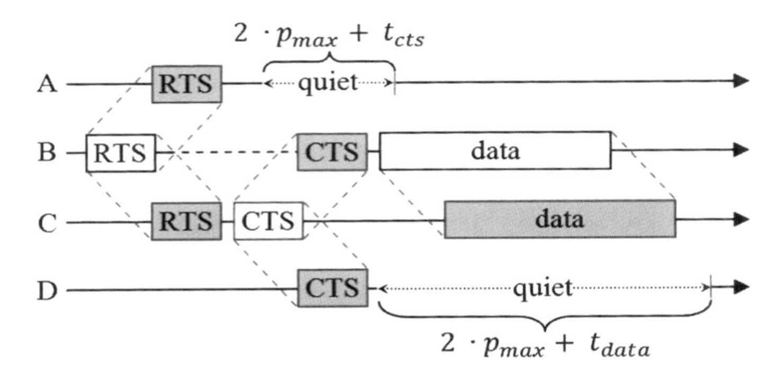

Fig. 3.12 MACA reservation handshake scheme

In the following sections we introduce the major handshake-based protocols using control packets or tones.

B.2.1.1 MACA

The *Multiple Access Collision Avoidance* (MACA) protocol, proposed by Karn [32], was the first approach of a handshake algorithm to reserve the channel, as shown in Fig. 3.12. The transmitter sends instead of the data packet a short *request-to-send* (RTS) if it wants to reserve the channel for its transmission. This RTS packet contains the length of the data packet which should be transmitted to define the length of the reserved slot. The receiver replies with a *clear-to-send* (CTS) packet which also contains the length of the data packet, if the channel is free and not reserved for other transmissions. After receiving the CTS packet, the transmitter starts immediately the transmission of the data packet. A node that overhears an RTS packet defers long enough to let the transmitter receive the corresponding CTS (twice maximum propagation delay plus transmission time of a CTS) and a node that overhears a CTS defers long enough to let the destination receive the data packet (twice maximum propagation delay plus the transmission time of the data packet, which is specified in the CTS).

If the channel at node B is already reserved, B ignores the RTS packet. Node A uses a timer that is started after sending the RTS packet, and if the timer runs out it repeats the channel request after a random backoff (a multiple of maximum propagation delay plus transmission time of an RTS), as shown in Fig. 3.13.

This implementation of the handshake solves the *hidden node problem* (Fig. 3.11a) but not as proposed the *exposed node problem* (Fig. 3.11b). The problem is that node C cannot receive the CTS packets from node D, because the CTS overlaps at C with B's ongoing transmission. Due to this two-way signalling needed for the reservation, node C is still exposed during B's transmission to A.

Additionally, a second *exposed node problem* is originated, as shown in Fig. 3.14. Without a handshake, node A can transmit to node B and in parallel node D can transmit to node C. Now D must first initiate a handshake and C must reply with a CTS, which would collide with the ongoing transmission of node A at

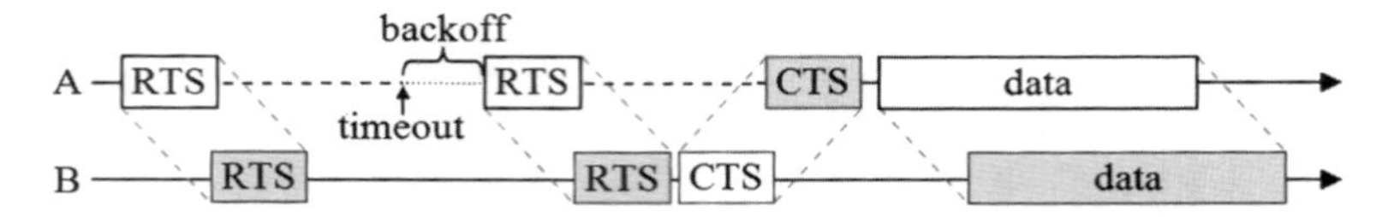

Fig. 3.13 MACA reservation repetition

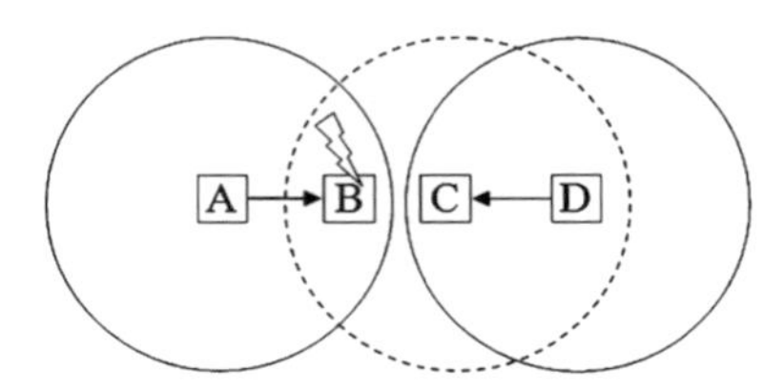

Fig. 3.14 Additional Exposed Node Problem in MACA

node B. C is aware of this, and therefore a channel reservation between nodes D and C cannot be done.

Furthermore, simultaneous handshakes can be a problem, especially in environments with high propagation delays like in underwater networks. For example, another channel reservation may arrive after a node already sent the CTS packet. Respectively, a node may start its transmission before overhearing a foreign CTS, resulting in a packet collision. Or a control packet may be lost and the simultaneous reservation is not heard. Therefore, total collision freeness cannot be guaranteed.

Analogous to ALOHA, MACA can also be extended with acknowledgment packets on the link layer, it was proposed by Bharghavan et al. [33] and is called *MACA Wireless* (MACAW). It was shown that retransmission at the link layer at each hop and not at the end-to-end connection increases the overall throughput in channels with low reliability, what is the case in underwater environments.

As described in the introduction of this section, a handshake only increases the throughput if data packets are long and there would be a lot of collisions without a preceding reservation. In *mobile Ad Hoc networks* (MANETs) it is shown that MACA mostly decreases the throughput instead increases the delay. Furthermore, additional control packets lead to a higher packet collision probability. Another problem in underwater networks is the high propagation delay, which extends the handshake duration of MACA significantly and a lot of bandwidth is wasted by waiting for control messages [30].

B.2.1.2 MACA-U

The media access protocol *MACA for Underwater* (MACA-U), proposed by Ng et al. [34], is an adaption of MACA for multi-hop underwater acoustic networks. They worked out an enhancement of the state transition rules of the original MACA which is designed for terrestrial environments. In MACA a node *waiting for CTS* (after sending an RTS) or *waiting for data* (after sending a CTS) transits into *quiet* state if it overhears any foreign RTS or CTS packet, which occurs more often in environments with high delays. Instead in MACA-U nodes ignore these packets and stay in *waiting for CTS* respectively *waiting for data* state. Only if a node receives a foreign CTS during *waiting for CTS* it transits into *quiet* state for twice the maximum propagation delay plus transmission time of the foreign data packet and retries the channel reservation at a later point in time. The second enhancement is extending the *quiet* state if necessary after overhearing a foreign RTS or CTS packet.

Another adaption is done in the packet forwarding strategy for multi-hop networks. MACA-U uses a second *First-In-First-Out* (FIFO) queue and separates data originated from the node itself and relay data. The *relay queue* has a higher priority and RTS packets corresponding to relay packets have an additional priority flag to enhance packet forwarding.

As a third adaption Ng et al. [34] tested different backoff algorithms besides the original *Binary Exponential Backoff* (BEB). They observed that a better throughput is achieved when the backoff algorithm assumes a drastic decrement like BEB, whereas linear decrease like *Multiplicative Increase Linearly Decrease* (MILD) reacts too slow to changing conditions.

Their simulation results show that MACA-U outperforms the original MACA by around 20% for the saturation throughput and is introduced as a better benchmark for comparison of new medium access protocols for underwater networks. They also compared MACA-U without and with carrier sensing (CS-MACA-U) which showed similar results in throughput.

B.2.1.3 PCAP

The *propagation-delay-tolerant collision avoidance protocol* (PCAP), proposed by Guo et al. [30], adapts the handshake of MACA for underwater acoustic networks. It tries to solve the problem of high signal propagation delays present in underwater networks. It is shown, that handshaking protocols become less efficient as the propagation delays increase, because of the long idle times during waiting for feedback. Handshaking showed poor throughput if the propagation delay is considerably larger than the transmission time of a data packet. In PCAP the transmission of a CTS is deferred, so that it reaches the transmitter after twice the maximum propagation delay after sending the RTS, as shown in Fig. 3.15.

The waiting time t_{wait} is given as:

$$t_{wait} = 2 \cdot \left(p_{\max} - p_{A,B}\right)$$

where p_{max} is the maximum propagation delay and $p_{A,B}$ the delay between node A and node B, which is calculated out of a transmission time t_{tx} included in the RTS and the receive time t_{rx} of the RTS at the receiver B:

$$p_{A,B} = t_{rx} - t_{tx}.$$

Due to the use of the absolute time difference $t_{rx} - t_{tx}$, PCAP needs time synchronization of all nodes. Time error introduced by clock drift should be negligible compared to the propagation delay.

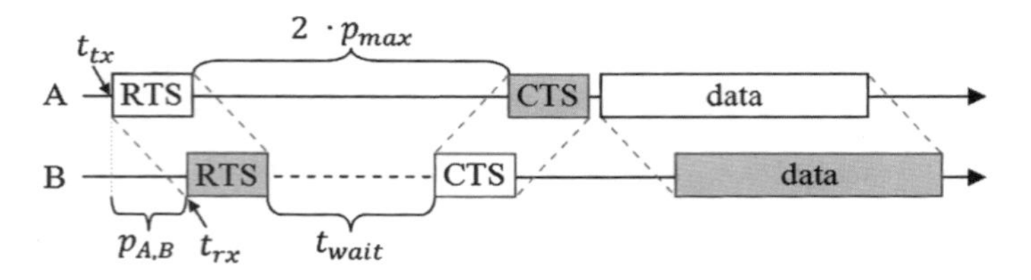

Fig. 3.15 PCAP reservation handshake scheme

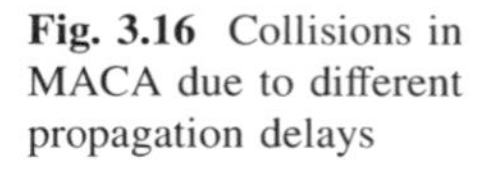
Fig. 3.16 Collisions in MACA due to different propagation delays

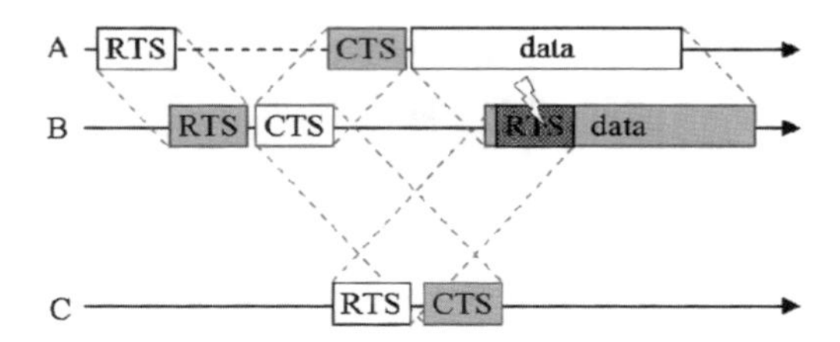

In PCAP the transmitter or neighbours can take other actions while waiting for the CTS, like transmitting another data frame or perform handshaking of the next data packet. The simulations in [30] showed significant improvements of the throughput (compared to other methods) if the propagation delays compared to the transmission times are long. The disadvantage of PCAP is the need of synchronization, and it is only suitable for delay tolerant applications due to the significantly higher handshake delay before each data transmission.

B.2.1.4 FAMA

In the *floor acquisition multiple access* (FAMA) protocol, proposed by Fullmer and Garcia [35], collisions due to simultaneous reservations are prevented by increasing the RTS and CTS packets. As mentioned, MACA cannot guarantee collision freeness, due to different propagation delays. For example this can happen, when one neighbour is near and one far and a CTS from a reservation arrives too late to prevent a third station from sending its own RTS, which will collide with the data packet, as shown in Fig. 3.16.

In FAMA this problem is solved by increasing the RTS and CTS transmission times. The length of an RTS is larger than the maximum channel propagation delay plus the transmit-to-receive turn-around time and any processing time for carrier detection. The length of a CTS is larger than the length of an RTS plus one maximum roundtrip time across the channel (twice the maximum channel propagation delay). The relationship of the size of the CTS to the RTS gives the CTS a so called *dominance* over the RTS in the channel. Even though a node starts the transmission of an RTS just before a CTS arrives and it is deaf during the transmission, it will hear the end of the CTS. Therefore the dominating CTS plays the role of a busy tone in FAMA and guarantees a collision free transmission of data packets.

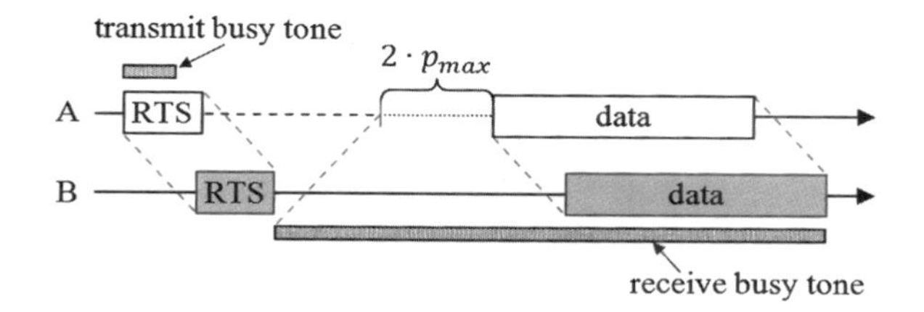

Fig. 3.17 DBTMA reservation handshake scheme

The increased length of the control packets results in a maximum handshake length of five times of the maximum propagation delay, three due to RTS and CTS transmissions and two due to packet propagation delay. In underwater networks this results in high packet delays and waste of energy due to the long transmission times of control packets. Therefore, Molins and Stojanovic proposed a slotted version of FAMA for underwater networks as described in the section on slotted access protocols (C).

B.2.1.5 DBTMA

In the *dual busy tone multiple access* (DBTMA), proposed by Deng et al. [31], the channel is divided into two subchannels, a data channel and a control channel. This protocol solves the exposed node problems, which are still present using MACA. Therefore, two control tones on a separate channel are included. A *transmit busy tone* is sent parallel to RTS packets to provide protection for the RTS packets to increase the probability of successful RTS reception at the intended receiver. The second *receive busy tone* replaces the CTS packet, which is send until the data packet is completely received, as shown in Fig. 3.17.

The transmitter A waits twice the maximum propagation delay after receiving the *receive busy tone* to allow all other nodes in range of the receiver B to abort possible RTS transmissions. This waiting time increases the handshake duration additionally and makes DBTMA unsuitable for underwater networks.

The *exposed node problem* (Fig. 3.11b) is solved, because node C can now receive the *receive busy tone* as CTS of the receiver node D, because it is on another channel and does not collide with the transmission of node B. The second *exposed node problem* (Fig. 3.14) is solved too; because node C can now reply with a CTS tone without interfering the transmission of node A and B.

Collision freeness cannot be guaranteed, because the RTS packet can still collide with data packets (Fig. 3.16) due to different propagation delays. The RTS itself cannot be replaced by a simple tone, because it must contain the data packet length and the destination node. Replacing the CTS by a simple *receive busy tone* does no longer allow to associate the tone to the corresponding RTS. Therefore, *receive busy tones* may confirm the wrong or more than one RTS packet resulting in collisions at the receiver.

The authors of DBTMA showed an enhancement compared to MACA, which may be useful for underwater networks too, but the control tones need additional

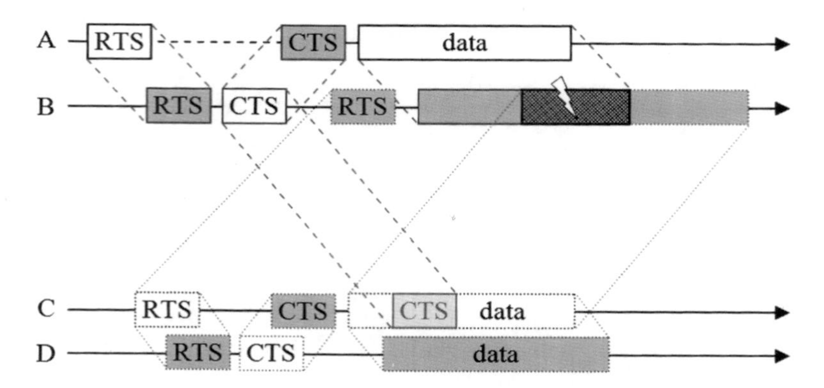

Fig. 3.18 Parallel handshakes with high propagation delays

energy and parallel receiving/transmitting of a control tone must be supported by the modems.

B.2.1.6 DACAP

The *Distance-Aware Collision Avoidance Protocol* (DACAP), proposed by Peleato and Stojanovic [36], is designed for underwater networks with high propagation delays. DACAP solves two problems arising out of parallel handshakes combined with long propagation delays. The first problem is already introduced in the FAMA section (B.2.1.4) and is shown in Fig. 3.16. To avoid collisions of data transmissions with foreign RTS packets, the transmitter defers its data transmission for t_{min} seconds after sending an RTS, where t_{min} is the minimum handshake length and predefined for all nodes. For networks wherein most node distances are close to the maximum transmission range, t_{min} needs to be nearly twice the maximum propagation delay. If some distances are shorter, t_{min} can be reduced and thereby the average handshake length. The second problem is shown in Fig. 3.18. Node A wants to transmit data to node B and initiates a handshake. In parallel, node C wants to transmit data to node D and initiates also a handshake. With the standard MACA RTS/CTS mechanism this can result in a collision of the data packet at node B if B receives the RTS from node C after already having sent the CTS to node A and node C starts the transmission before overhearing the CTS of node B.

To solve this problem, DACAP introduces a very short warning packet sent by the receiver if it overhears an RTS within $2 \cdot p_{max} - t_{min}$ after sending a CTS, where p_{max} is the maximum propagation delay, shown in Fig. 3.19. Because node A waits a certain time before beginning the data transmission, it can overhear the warning of node B and defer the transmission. Also node C defers its transmission because it overhears the CTS from node B during the waiting period. After a random backoff node A and node C reinitiate the handshake.

In [36], two versions of DACAP are introduced, with and without additional acknowledgements. The simulations show a higher throughput of DACAP than

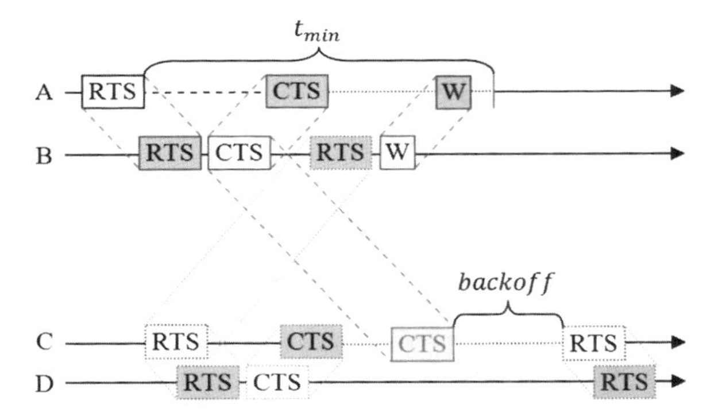

Fig. 3.19 The waiting and warning mechanism from DACAP

Slotted FAMA and in some cases than ALOHA. The disadvantage of DACAP is that t_{min} has to be configured in advance and is not adaptive to changing network topologies. It has to be chosen as trade-off between longer handshake/idle times and collision probability. The enhancement of using warning packets additional to the waiting periods is not separately shown and therefore unclear. In the scenario shown in Fig. 3.19 the warning packet of node B leads to the deferment of both data packets; leaving out the warning defers only one transmission and would be more effective in this case.

B.2.2 Contention-based Reservation

With contention-based reservation, in advance of each data transmission all nodes with data ready to send compete with each other for channel access in so called contention rounds. Each node that wants to reserve the channel for a data transmission, signals this to all neighbours by transmitting a control tone or message when the channel is free of activities. If there is more than one node contending for the channel, each node chooses a backoff (e.g., using a random generator) and the next contention round is started. If a node overhears a reservation of a competitor during this backoff period, it has lost the contention and must wait. Due to the propagation delay there can be still more than one contender left. Therefore, this will be repeated until there is only one node left. Now, the winner is allowed to transmit its data packet. After the transmission has finished, all nodes with data to send compete for channel access again.

B.2.2.1 Tone-Lohi

The *Tone-Lohi* (T-Lohi) protocol, proposed by Syed et al. [37], uses tones during contention rounds to reserve the channel. Contender detection during a contention round comes from listening to the channel after sending the short reservation tone.

Since underwater acoustic modems are typically half-duplex, a node that is transmitting a reservation tone cannot receive another tone at the same time. Therefore, nodes become *deaf* to simultaneous reservations if the time difference between the transmissions ($t_A - t_B$) of node A and node B minus the propagation delay ($p_{A,B}$) is smaller than the time needed to detect a reservation tone (T_{detect}):

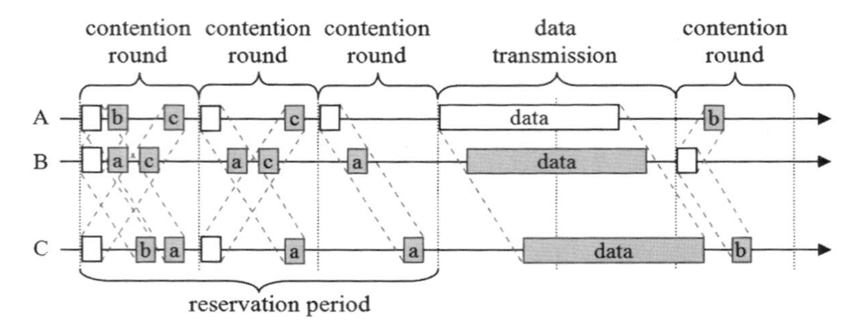

Fig. 3.20 The synchronized T-Lohi reservation contention-based scheme

$$\left||t_A - t_B| - p_{A,B}\right| < T_{detect}$$

In this case one node detects the channel erroneously as free and starts data transmission before the reservation is finished. This problem is mitigated if there are more than two contenders with different propagation delays.

T-Lohi takes advantage of the high propagation delays present in underwater networks to count the number of contenders during a contention round. Because the tones are short and the propagation delay is relatively high, the tones of different contenders mostly arrive in sequence and not in parallel, as shown in Fig. 3.20.

Even if tones collide, the presence of some contenders can still be determined if tones are not destructive. This contender counting allows T-Lohi to react dynamically to the actual traffic by adapting the backoffs during the contention rounds to the number of contenders.

Syed et al. [37] proposed three flavours of T-Lohi with different reservation mechanisms. A synchronized version (ST-Lohi) and two unsynchronized versions, the *conservative Unsynchronized T-Lohi* (cUT-Lohi) and the *aggressive Unsynchronized T-Lohi* (aUT-Lohi):

a) Synchronized Tone-Lohi (ST-Lohi):

The synchronized flavour of T-Lohi needs global time synchronization of all nodes, because the transmission start times are synchronized. Therefore, ST-Lohi is actually a slotted protocol of category C. It is described at this point, because ST-Lohi is the foundation for the unsynchronized flavours. Slots are as long as a contention round whose length is defined as the maximum propagation length plus the tone detection time to guarantee the recognition of all contenders during a contention round:

$$CR_{ST} = p_{max} + T_{detect}$$

If a node has data to send, it must wait until the beginning of the next contention round, where it is allowed to start channel reservation, if the channel is not already reserved or there is an ongoing reservation. Otherwise they have to wait until the

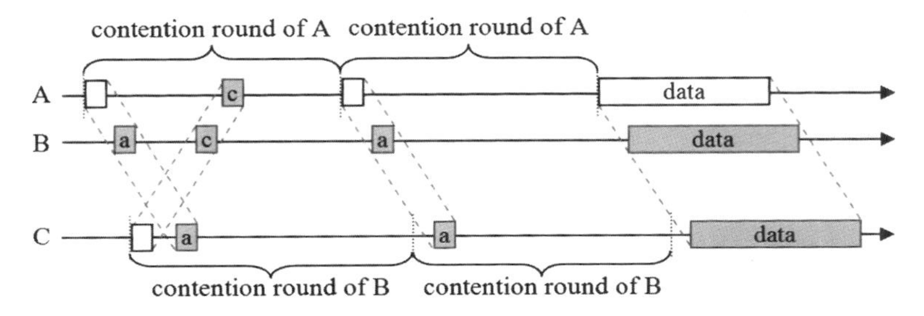

Fig. 3.21 The conservative unsynchronized T-Lohi reservation scheme

channel is free. Each node that has data to send transmits a short reservation tone and listens until the end of the contention round to count the number of contenders, as shown in Fig. 3.20.

In ST-Lohi the nodes should not be too close, because contenders are only detected if the propagation delay is equal or greater than the time needed to detect a reservation tone:

$$p_{A,B} \geq T_{detect}$$

In Fig. 3.20, node A, node B, and node C want to reserve the channel. After detecting and counting the contenders each node draws its own backoff ω uniformly from $[0, CRC]$, where CRC is the contender count. Each node must wait this ω contention rounds before it is allowed trying to reserve the channel again, assumed that no other contender did choose a shorter backoff. Consequently, the number of contenders is reduced each round until there is a winner, like node A in Fig. 3.20. After node A did not sense any contender in round three, it is allowed to start the data transmission at the beginning of the next round. Due to the dynamic adaption of the backoff this mechanism converges in very few rounds; even for a high number of contenders ST-Lohi needs just three or four contention rounds.

Additionally, the authors of T-Lohi pointed out that in underwater networks there exists a spatial unfairness due to the high propagation delays. In Fig. 3.20 node C detects the channel as free, after the data transmission of node A, later than a nearer node B, which therefore can reserve the channel one slot earlier. To reduce this spatial unfairness, in ST-Lohi nodes have to wait an additional time depending on the distance to the last transmitter.

b) Unsynchronized Tone-Lohi (UT-Lohi):

In unsynchronized Tone-Lohi (UT-Lohi), nodes can contend for the channel any time they know the channel is not busy, as shown in Fig. 3.21. To enable the counting of all contenders the contention round is doubled in the *conservative unsynchronized T-Lohi* (cUT-Lohi):

$$CR_{cUT} = 2 \cdot CR_{ST} = 2 \cdot p_{max} + 2 \cdot T_{detect}$$

This additional time is needed if a contender is as far away as the maximum transmission range and it transmits its tone just before detecting the preceding

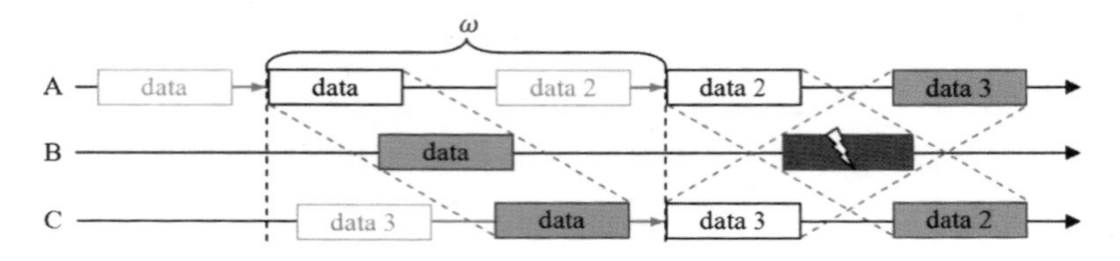

Fig. 3.22 Slotted ALOHA channel access scheme

tone. Due to reasons of deafness conditions even the doubled contention round does not guarantee a successful reservation, but reduces collisions to a minimum.

To reduce the reservation period another flavour of T-Lohi is introduced, called *aggressive unsynchronized T-Lohi* (aUT-Lohi) where the contention period is the same as in ST-Lohi:

$$CR_{aUT} = CR_{ST} = CR_{cUT}/2 = p_{max} + T_{detect}$$

Consequently, aUT-Lohi reduces the idle times but conversely increases the propability of collisions due to premature reservations. The simulations in [37] show a higher channel utilization of aUT-Lohi compared with cUT-Lohi.

The advantage of T-Lohi is the very low energy consumption, because the reservation scheme reduces data packet collisions and the reservation tones can be used as special wake-up tones, which additionally reduces the energy consumption during channel listening. The disadvantage of T-Lohi is that it is designed for short-ranged fully connected networks. The reservation mechanism does not regard hidden stations, which are considerably present in multi-hop networks. Simulations [38] showed reduced performance of all T-Lohi flavors in such environments.

Another problem is the dependency between the contention round duration and the maximum propagation delay; so high transmission ranges reduce the channel utilization. Even with perfect scheduling T-Lohi reaches a maximum throughput of:

$$Utilization = \frac{t_{data}}{t_{data} + CR}$$

where t_{data} is the packet transmission time and CR the contention round duration, which is a multiple of the maximum propagation delay, depending on the used flavour of T-Lohi. T-Lohi is not considered suitable for wide-range underwater multi-hop networks.

C. Slotted Access

In slotted access protocols the time is divided in slots of length ω and a transmission is deferred to the beginning of the next time slot, like shown in Fig. 3.22. Typically, random access protocols of category B are extended to slotted access. Restricting packet transmission to predetermined time slots decreases the probability of packet collisions due to partial overlapping. In slotted access packet collisions can only occur if two nodes select the same time slot. But slotting increases the idle times, resulting in higher packet delays and lower channel utilization by reducing the collision probability. Like TDMA, in *slotted* protocols all nodes must be synchronised and the slot length ω is proportional to the propagation length, which can be a problem in wide-ranged underwater networks.

C.1 Slotted Aloha

In the slotted version of ALOHA (*slotted ALOHA*), proposed by Roberts [39], the design principle of random access without reservation of pure ALOHA is maintained. Only the channel access, and therefore the data transmission, is deferred to the beginning of the next time slot, shown in Fig. 3.22. Since a slot is not assigned to a single node like in TDMA, collisions are still possible if multiple nodes select the same time slot. But even if two nodes select the same slot, the simultaneous transmissions do not have to collide. For example, if node A and node C in Fig. 3.22 transmit in parallel, they receive the data packets of the other node due to the propagation delay. But at the middle node B they collide and cannot be decoded.

C.2 Slotted FAMA

The slotted version of the *floor acquisition multiple access* (FAMA) protocol, proposed by Molins and Stojanovic [40], does not need the long transmission times of RTS and CTS packets as the original FAMA [35] does. The increased lengths of control packets are introduced in FAMA to eliminate collisions of data packets like in MACA due to different packet delays, as shown in Fig. 3.16. To prevent other nodes from sending when data is being sent, in FAMA the length of the RTS and CTS packets are larger than the maximum propagation delay, which is not a suitable solution for underwater networks. *Slotted FAMA* uses time slotting which eliminates the need for excessively long control packets, thus providing savings in energy. The slot length is set to the maximum propagation delay plus the transmission time of a CTS. Due to the fact that a node must wait until the beginning of the next time slot, it overhears the CTS from a competitor before starting its own handshake, as shown in Fig. 3.23. In consequence, *Slotted FAMA* guarantees a collision free transmission of data packets; only the control packets of simultaneous handshakes can collide. Note that the data packets are allowed to be longer than one slot length as long as the required transmission time was included in the control packets of the previous handshake. Therefore, it is possible to transmit multiple packets at once, a so called *packet train*, with the need of only one prior handshake.

In *Slotted FAMA*, the reduction in control packet lengths are an improvement of the original FAMA in underwater environments, especially if the maximum transmission range and the resulting propagation delay is high. Even though the energy efficiency is good, the channel utilization is still low in long-range networks; equally the packet delay is still high, because the handshake does still need two slots and therefore twice the maximum propagation delay.

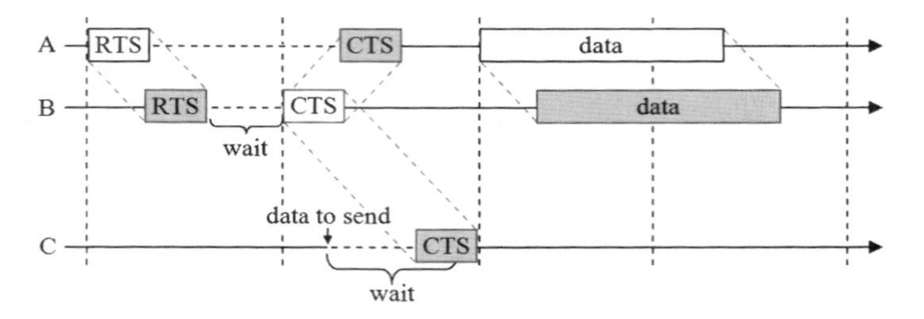

Fig. 3.23 Channel access scheme of slotted FAMA

3.3.3 Medium Access Cooperation with Game Theory

Using time based medium access in ad-hoc networks, collisions are always possible. Standard procedures to avoid collisions lead to a high administrative effort. Game theory offers a promising set of tools to analytically model ad-hoc networks and a design of incentive mechanisms to design robust protocols dealing with selfish behavior [41, 42]. "Intelligence Game theory can be defined as the study of mathematical models of conflict and cooperation between intelligent rational decision-makers. Game theory provides general mathematical techniques for analyzing situations in which two or more individuals make decisions that will influence on another's welfare" [43].

With game theory, the probability of a packet transmission from a neighbour should be estimated. The objective of each node is to maximize the individual throughput by reducing the collisions [44]. The probability that a neighbour will transmit in the next time may be guessed from overheard traffic, for example if a node overhears a packet that its neighbour has to forward. [45, 46]. The game theory approach can also be applied to routing algorithms [47, 48].

3.3.4 Discussion of Existing Time Based Multiple Access Technologies

If propagation delays are small relative to message lengths, protocols utilizing channel reservation (category B.2) or time division multiplexing (category A) may be the most efficient choice.

But in applications with decentralized network topology and high propagation delays relative to message length, time division multiplexing protocols (category A) are not suitable. The channel would not be used efficiently, because the required guard times between each slot have the order of the maximum propagation delay. Additionally node synchronization would be needed, which cannot be easily assumed in an ad hoc network. These arguments against protocols of category A are also true for protocols of category C with slotted access.

When there are high propagation delays relative to message length, the protocols of category B.2 with preceding channel reservation will also achieve only low channel utilizations due to the long handshake durations of twice the maximum propagation delay. Therefore, the nodes will spend the most part of the time waiting for confirmations of channel reservations. To get acceptable channel utilization compared to random access protocols, the packet transmission times and consequently the probability of collisions must be very high.

As a consequence, random access protocols without channel reservation of category B.1, e.g. ALOHA, seems to be the best solution when propagation delays are long relative to message length. They are detached from the maximum propagation delay but have the problem of collisions. To reduce collisions to a minimum, the packet sizes and therefore the transmission times should be as small as possible. This is reinforced by simulations [38] with the *World Ocean Simulation System* (WOSS) [49], where ALOHA is compared with T-Lohi and DCAP. The lower overhead and shorter channel access delay of ALOHA yields better performance simulating small packet sizes. If long packet sizes are used, one has to be aware that this will come at the cost of packet loss due to collisions.

Random access strategies may be combined with effective flooding algorithms, which allows the use of additional network coding strategies (Sect. 4.4.5) to deal with the bad channel conditions, and exploit the time diversity originated from the high propagation delays with collaborative beamforming approaches (Sect. 4.4.6). This can be combined with the utilization of check sums and reliabilities from the physical layer. Additionally, game theory strategies may reduce the collision probability in networks with a shared medium like in the acoustic underwater communication [50] as described in Sect. 3.3.3.

3.4 Combination of Different Multiple Access Schemes

It is also possible to combine different multiple access schemes. For example, one can combine time-based ALOHA multiple access with FDMA or CDMA, called *multichannel ALOHA* [51]. In the case of multichannnel ALOHA with FDMA, the total frequency band can be split into several bands, and when a transmitter has data to send it can send it on one of the frequency bands. The choice of frequency band can be random or determined by some algorithm.

One challenge in multichannel ALOHA is that the intended receiver must be listening to the correct channel. If sufficient processing power is available at the receiver, it could simultaneously listen for transmissions on all channels. If this is not feasible, a common control channel has to be used: When a transmitter has data to send, it selects a channel for communication and transmits a short burst on the control channel telling the intended receiver to listen at the correct channel.

Zhou et al. [52, 53] have investigated multichannel MAC methods for underwater acoustic networks. They compare multichannel ALOHA with multichannel

RTS/CTS, and in [52] they present various plots comparing the two approaches for different traffic parameters and number of channels. The two approaches compare similarly as in single-channel systems with respect to packet lengths, traffic load, and propagation delay. But in their simulations the performance of multichannel RTS/CTS is improved as the number of channels increases, while the performance of multichannel ALOHA is relatively independent of the number of channels. In [53] they discuss additional hidden terminal problems that emerge in multichannel systems, and propose to augment multichannel RTS/CTS with control tones to alleviate this.

As mentioned in Sect. 3.2.3, Pompili et al. [11] have also proposed and investigated a CDMA-based multichannel ALOHA scheme.

Note that the challenges in CDMA and FDMA related to possibly large difference in simultaneously received signals on different channels (see Sects. 3.1 and 3.2) also have to be considered in multichannel ALOHA or multichannel RTS/CTS.

References

1. Sozer EM, Stojanovic M, Proakis JG (2000) Underwater acoustic networks. IEEE J Ocean Eng 25(1):72–83
2. Akyildiz IF, Pompili D, Melodia T (2005) Underwater acoustic sensor networks: research challenges. Ad Hoc Netw 3:257–279
3. Jarvis S, Janiesch R, Fitzpatrick K, Morrissey R (1997) Results from recent sea trials of the underwater digital acoustic telemetry system. In: Proceedings of 7th international conference on electronic engineering in oceanography, Southampton, UK, Halifax, Nova Scotia, pp 186–192
4. Rice J, Creber B, Fletcher C, Baxley P, Rogers K, McDonald K, Rees D, Wolf M, Merriam S, Mehio R, Proakis J, Scussel K, Porta D, Baker J, Hardiman J, Green D (2000) Evolution of Seaweb underwater acoustic networking. In: Proceedings of MTS/IEEE Oceans, vol 3, Providence, RI, USA, pp 2007–2017
5. Aliesawi S, Tsimenidis CC, Sharif BS, Johnston M, Hinton OR (2010) Adaptive multiuser detection with decision feedback equalization based IDMA systems on underwater acoustic channels. In: Proceedings of European Conference on Underwater Acoustics, ECUA 2000, Istanbul, Turkey, pp 1674–1679
6. Passerieux J-M, Robert C, Fischer S, van Walree P, Robert M, Wilmink E, Nelisse M, Adams A, Açar G, Coatelan S (2004) ACME project: final report. Technical report TUS SAS 04/S/EGS/NC/064-JMP
7. van Walree PA (2003) Second ACME sea trials: analysis of the acoustic channel. Technical report project deliverable, Jan 2003
8. Açar G, Adams AE (2006) ACMENet: an underwater acoustic sensor network protocol for real-time environmental monitoring in coastal areas. IEEE Proc Radar Sonar Navig 153(4):365–380
9. Xie GG, Gibson JA (2000) A networking protocol for underwater acoustic networks. Technical report TR-CS-00-02
10. Hwee-Pink Tan, Winston KG. Seah, Linda Doyle. (2007) A multi-hop ARQ protocol for underwater acoustic networks. In: Proceedings of IEEE Oceans 2007 Europe, Aberdeen, UK
11. Pompili D, Melodia T, Akyildiz IF (2009) A CDMA based medium access control for underwater acoustic sensor networks. IEEE Trans Wireless Commun 8(4):1899–1509

12. Watfa MK, Selman S, Denkilkian H (2010) UW-MAC: an underwater sensor network MAC protocol. Int J Commun Syst 23(4):485–506
13. Falconer DD, Adachi F, Gudmundson B (1995) Time division multiple access methods for wireless personal communications. IEEE Commun Mag 33(1):50–57
14. Garg VK (2007) Wireless communications and networking. The Morgan Kaufmann series in networking, illustrated edition 2007
15. Dongfeng Z, Bihaiy L, Sumin Z (1996) Analysis of a slotted access channel with average cycle method. International conference on publication communication technology proceedings (ICCT'96), vol 2, pp 1080–1083
16. Liu J, Zhou Z, Peng Z, Cui JH (2010) Mobi-sync: Efficient time synchronization for mobile underwater sensor networks. Technical report UbiNet-TR10-01, UCONN CSE
17. Proakis JG, Sozer EM, Rice JA, Stojanovic M (2001) Shallow water acoustic networks. IEEE Commun Mag 39(11):114–119
18. Catipovic J, Brady D, Etchenmendy S (1993) Development of underwater acoustic modems and networks. Oceanography 6:112–119
19. Freitag L, Grund M, von Alt C, Stokey R, Austin T (2005) A shallow water acoustic network for mine countermeasures operations with autonomous underwater vehicles. In: Proceedings of Underwater Defense Technology (UDT), Amsterdam, Netherlands
20. Chlamtac I, Kutten S (1985) A spatial reuse TDMA/FDMA for mobile multi-hop radio networks. In: Proceedings of IEEE INFOCOM'85, vol 1, Washington DC, USA, pp 389–394
21. Garey MR, Johnson DS (1990) Computers and intractability; A guide to the theory of NP-completeness. WH Freeman & Co., New York
22. Chlamtac I, Pinter SS (1987) Distributed nodes organization algorithm for channel access in a multihop dynamic radio network. IEEE Trans Comput 36:728–737
23. Colin YM. Chan, Mehul Motani (2007) An integrated energy efficient data retrieval protocol for underwater delay tolerant networks. In: Proceedings of IEEE oceans 2007 Europe, Aberdeen, UK
24. Baker D, Ephremides A (1981) The architectural organization of a mobile radio network via a distributed algorithm. IEEE Commun Trans 29(11):1694–1701
25. Fischer MJ (1983) The consensus problem in unreliable distributed systems (A brief survey). In: Proceedings of the 1983 international FCT-conference on fundamentals of computation theory, Springer, London, 127–140
26. Kanzaki A, Uemukai T, Hara T, Nishio S (2003) Dynamic TDMA slot assignment in ad hoc networks. In: International conference on advanced information networking and applications, Xi'an, China, pp 330–335
27. Apostolas C, Tafazolli R, Evans BG (1995) Wireless ATM LAN. In: IEEE International Symposium on Personal, Indoor and Mobile Radio Communications, PIMRC, vol 2, Toronto, ON, Canada, pp 773–777
28. Apostolas C, Tafazolli R, Evans BG (1996) Comparison between elimination yield non pre-emptive priority multiple access (EY-NPMA) and dynamic TDMA (D-TDMA). In: IEEE International Symposium on Personal, Indoor and Mobile Radio Communications, PIMRC, vol 2, Taipei, Taiwan, pp 663–667
29. Abramson N (1985) Development of the ALOHANET. IEEE Trans Inf Theory 31(2):119–123
30. Guo X, Frater MR, Ryan MJ (2006) A propagation-delay-tolerant collision avoidance protocol for underwater acoustic sensor networks. In: Proceedings of IEEE Oceans 2006 Pacific, Singapore
31. Deng J, Haas ZJ (1998) Dual busy tone multiple access (DBTMA): a new medium access control for packet radio networks. In: Proceedings of international conference on Universal Personal Communications, ICUPC '98, vol 2, Florence, Italy, pp 973–977
32. Karn P (1990) MACA—a new channel access method for packet radio. In: ARRL/CRRL Amateur radio 9th computer networking conference, London, ON, Canada, pp 134–140
33. Bharghavan V, Demers A, Shenker S, Zhang L (1994) MACAW: a media access protocol for wireless LANs. In: Proceedings of ACM SIGCOMM, London, UK, 1994, pp 212–225

34. Ng H- H, Soh WS, Motani M (2008) MACA-U: a media access protocol for underwater acoustic networks. In: Proceedings of the Global telecommunications conference, IEEE GLOBECOM 2008, New Orleans, LO, USA
35. Fullmer CL, Garcia-Luna-Aceves JJ (1995) Floor acquisition multiple access (FAMA) for packet-radio networks. In: ACM SIGCOMM computer communication review vol 25, pp 262–273
36. Peleato B, Stojanovic M (2007) Distance aware collision avoidance protocol for ad-hoc underwater acoustic sensor networks. IEEE Commun Lett 11(12):1025–1027
37. Syed AA, Ye W, T-Lohi JH (2007) A new class of MAC protocols for underwater acoustic sensor networks. Technical report ISI-TR-638, USC/Information Sciences Institute
38. Guerra F, Casari P, Zorzi M (2009) A performance comparison of MAC protocols for underwater networks using a realistic channel simulator. In: IEEE Oceans'09, Biloxi, MS, USA
39. Roberts LG (1975) ALOHA packet system with and without slots and capture. SIGCOMM Comput Commun Rev 5(2):28–42
40. Molins M, Stojanovic M (2006) Slotted FAMA: a MAC protocol for underwater acoustic networks. In: IEEE OCEANS'06, Singapore, pp 16–19
41. Felegyhazi M, Cagalj M, Hubaux J-P (2009) Efficient MAC in cognitive radio systems: a game-theoretic approach. Wireless Commun IEEE Trans 8(4):1984–1995
42. Urpi A, Bonuccelli M, Giodano S (2003) Modelling cooperation in mobile ad hoc networks: a formal description of selfishness. In: Proceedings of modeling and optimization in mobile, ad hoc and wireless networks, Sophia Antipolis, France, pp 3–5
43. Myerson RB (1991) Game theory: analysis of conflict, Harvard University Press, Cambridge p 568
44. Buttyan L, Hubaux J-P (2001) Nuglets: a virtual currency to stimulate cooperation in self-organized mobile ad hoc networks. In: Technical report DSC/2001/001, Swiss federal institute of technology—Lausanne, Department of Communication Systems 2001
45. Felegyhazi M, Buttyan L, Hubaux J-P (2003) Equilibrium analysis of packet forwarding strategies in wireless ad hoc networks—the static case. In: Proceedings of PWC 2003 personal wireless communications, Venice, Italy, pp 3–25
46. Felegyhazi M, Hubaux J-P, Buttyan L (2006) Nash equilibria of packet forwarding strategies in wireless ad hoc networks. IEEE Trans Mob Comput 5(5):463–476
47. Marti S, Giuli TJ, Lai K, Baker M (2000) Mitigating routing misbehavior in mobile ad hoc networks. In: Proceedings of international conference on mobile computing and networking, ACM, Boston, MA, USA, pp 255–265
48. Orda A, Rom R, Shimkin N (1993) Competitive routing in multi-user communication networks. IEEE/ACM Trans Netw 1:510–521
49. Guerra F, Casari P, Zorzi M (2009) World ocean simulation system (WOSS): a simulation tool for underwater networks with realistic propagation modeling. In: Proceedings of WUWNet 2009, Berkeley, CA, USA
50. Ong CW (2008) A discovery process for initializing ad hoc underwater acoustic networks. Master's thesis, Naval Postgraduate School, Monterey, CA, USA
51. Pountourakis IE, Sykas ED (1992) Analysis, stability and optimization of Aloha-type protocols for multichannel networks. Comput Commun 15(10):619–629
52. Zhou Z, Peng Z, Cui J-H, Shi Z (2008) Analyzing multi-channel MAC protocols for underwater acoustic sensor networks. Technical report UbiNet-TR08-02, Computer Science & Engineering, University of Connecticut
53. Zhou Z, Peng Z, Cui J-H, Jiang Z (2009) Handling triple hidden terminal problems for multi-channel MAC in long-delay underwater sensor networks. Technical Report UbiNet-TR09-02, Computer Science and Engineering, University of Connecticut

Chapter 4
Logical Link Layer Topics

4.1 Scope

In this chapter, we describe methods to ensure reliable information delivery to higher layers at the sink(s), while keeping overhead, retransmissions, and discarded information as low as possible. The basic mechanism to achieve this is typically ARQ (automatic repeat request), but many variants and additional mechanisms can be used.

One important contributor to ensuring low error rates is forward error correction (FEC). FEC is however considered part of the physical layer, and not investigated in detail in this report. FEC reduces the error rates in the physical layer, such that link layer protocols have a better chance of being successful.

Some MAC protocols are tightly coupled with ARQ. This is particularly the case for MAC protocols using an RTS/CTS exchange to reserve the channel. They are often referred to as RTS/CTS/DATA/ACK exchange, where DATA/ACK is the ARQ mechanism. In these cases, the present chapter describes variants of the applied ARQ protocol, while MAC is described in the previous chapter.

In multinode networks, the link layer performance can be improved by cooperation between several nodes, instead of considering each node-to-node link separately and leave the rest to routing protocols in the network layer. This is discussed in Sect. 4.4. In these cases, the borderline between link layer and network layer may become unclear.

4.2 ARQ

ARQ is the basic mechanism to ensure that no erroneous frames are delivered to higher layers. In this context, each "frame" consists of a number n data bits and a cyclic redundancy check (CRC) of p bits, where $n + p < 2^{p-1}$. A typical choice is

R. Otnes et al., *Underwater Acoustic Networking Techniques*,
SpringerBriefs in Electrical and Computer Engineering,
DOI: 10.1007/978-3-642-25224-2_4, © The Author(s) 2012

$p = 16$. CRC codes have good error detection capability (but no error correction capability). The probability of undetected error is below 2^{-p} for a p-bit CRC, as long as n is not too short (n should be above 200 for a 16-bit CRC) [1]. Note that this reference assumes uncorrelated bit errors, which is not the case if a FEC is applied at the PHY layer. The recommended range of frame sizes for a 16-bit CRC is from 10 bytes to 4 kbytes.

In some applications 32-bit CRC may be considered, in order to reduce the residual error rate. This depends on the potential effect an error may lead to and whether there are error detecting mechanisms at higher layers.

The usual assumption is that if the CRC check passes, the frame has no errors. Then, the receiver sends an ACK (acknowledgment) message back to the transmitter. If the CRC check fails, the receiver sends a NAK (negative acknowledgment), such that the transmitter knows that the frame was received in error. If the transmitter does not receive an ACK message within a defined time-out time, it also assumes that the frame was lost, although in reality it may only have been the ACK message that was lost. Generally, control messages such as ACK/NAK should be sent with a more robust physical layer waveform than data frames, to avoid unnecessary retransmissions.

If the transmitter receives a NAK message or does not receive an ACK message, it retransmits the frame. If the frame is not successfully received within a defined maximum number of retransmissions, the transmitter gives up, and the frame is lost.

Note that a more robust physical layer will reduce the need for retransmissions. But still, some kind of ARQ mechanism is required at the link layer to ensure reliable delivery of the information.

4.2.1 Stop-and-Wait and Go-Back-N ARQ

The simplest ARQ variant is "stop-and-wait". This means that after transmitting one frame, the transmitter waits for ACK or NAK from the receiver before making its next transmission (retransmission or new frame). This is very inefficient in underwater communications due to high round-trip delay between transmitter and receiver, which means that an unnecessarily large fraction of the time will be spent waiting for acknowledgments.

A slightly more sophisticated ARQ variant is "go-back-N". This means that the transmitter sends several enumerated frames consecutively in one transmission, and in case of errors the receiver feeds back the sequence number of *the first* frame that was received in error. As an example, the transmitter sends six frames enumerated 1–6. Frames 3 and 5 are received in error, and the receiver feeds back that the first error was in frame 3. On its next transmission, the transmitter then goes back to frame 3 and continues from there, such that its next six frames to send would be number 3–8. Note that in this example, frame 4 and 6 are retransmitted even though they have already been correctly received. Go-back-N ARQ therefore

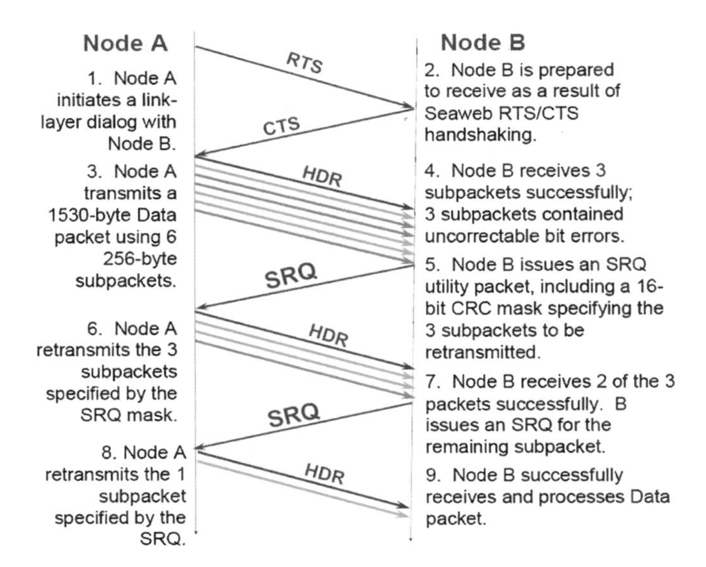

Fig. 4.1 Illustration of selective repeat ARQ as implemented in Seaweb. From [4] with permission. In the terminology of the present report, SRQ corresponds to selective repeat ARQ, and subpacket corresponds to frame. HDR means header

wastes transmission energy and time compared to selective repeat ARQ (Sect. 4.2.2). The benefit of go-back-N compared to selective repeat is that the buffering requirement on the receiver side is smaller.

We have not found any references recommending the use of stop-and-wait ARQ or go-back-N ARQ in underwater communications systems, at least not when compared to selective repeat ARQ.

4.2.2 Selective Repeat ARQ

In selective repeat ARQ, the transmitter sends several enumerated frames consecutively in one transmission, and in case of errors the receiver feeds back the sequence number of *all* frames that were received in error. On its next transmission, the transmitter then retransmits only the frames received in error, possibly together with new frames. With this approach, the time wasted waiting for ACK is reduced compared to stop-and-wait ARQ, and the time and energy wasted on unnecessary retransmissions is reduced compared to go-back-N ARQ.

Selective repeat ARQ is used in the US Seaweb underwater communications system [2–4], where it is called SRQ. The SRQ protocol used in Seaweb is illustrated in Fig. 4.1. In this implementation, the retransmitted frames are not concatenated with new frames. The header is 9 bytes, the frame size is 256 bytes ($n = 2048$ bits), and a 16-bit CRC is appended to each frame [4].

Go-back-N ARQ and selective repeat ARQ would perform even better with full-duplex (which is difficult under water, see Sect. 2.2) than with half-duplex. In [5], the half-duplex version of selective repeat ARQ is considered to be a variant of stop-and-wait ARQ, since the half-duplex operation still forces the transmitter to stop and wait for acknowledgments after some frames have been transmitted.

The optimal frame size and number of frames to send in each transmission depends on the bit error rate and on the propagation delay. This is analyzed in [5], where it is proposed to adapt these parameters to channel conditions and propagation delay, and also to adapt the time-out values to the propagation delay.

4.3 Hybrid ARQ

4.3.1 Type I Hybrid ARQ

"Hybrid ARQ" refers to combining ARQ (error *detection* and retransmission) with the error *correction* capabilities of FEC. Its simplest form called type I hybrid ARQ just refers to using FEC at the physical layer and ARQ at the link layer. Since FEC has to be used at the physical layer in any practical underwater communication system, all the ARQ variants described above will actually be type-I hybrid ARQ when applied to such systems.

4.3.2 Type II Hybrid ARQ

Type II hybrid ARQ is also referred to as incremental redundancy ARQ. It relies on a FEC code at the physical layer which is able to provide different amounts of redundancy, i.e., has outputs corresponding to different code rates. As a simple example, assume that the FEC is a convolutional code with four different generating polynomials, such that the code rate is 1/4 if all four shift register outputs are used. On the first transmission of a frame, the transmitter only sends the output from two shift registers (code rate 1/2). If the frame is not received and decoded correctly, the transmitter sends a new frame with the output from the other two shift registers. The receiver then combines the two received frames with a decoder for the overall rate 1/4 code. More generally, the incremental redundancy required for type II hybrid ARQ can be realized by applying different puncturing patterns to a code, or by applying fountain codes (Sect. 4.3.3).

The benefit of type II hybrid ARQ is that the overall FEC code combined from several transmitted frames with different redundancy information will have better performance than just repeating the same transmission several times.

Type II hybrid ARQ for underwater acoustic communications is studied in [6], where its performance is assessed based on a Markov model for SNR variations derived from actual channel measurements.

For High Frequency (3–30 MHz) radio communications, which is to considered to be among the most difficult RF propagation channels, data link protocols using burst waveforms and hybrid type II ARQ have been standardized [7, 8]. The protocols are called LDL and HDL (Low-Rate and High-Rate Data Link), and the incremental redundancy is based on convolutional codes with up to four shift register outputs. An extension called HDL+ [9], which uses adaptive data rate and has higher rates available, has also been proposed for standardization. One might consider adopting ideas from these protocols for underwater acoustic links.

4.3.3 Fountain Codes (Rateless Codes)

In challenging and time-varying communication environments such as underwater acoustic channels, the channel capacity and hence the optimal data rate will vary with time. Therefore, varying the information data rate with time will improve the throughput of the system. This can be achieved using adaptive modulation (Sect. 2.3), which requires the receiver to feed back information on the channel quality ("channel state information"). Note that in the physical layer context "channel state information" may refer to knowledge of the channel impulse response, but in the present context it is most often quantized information on SNR only.

The transmitter can use the channel state information fed back from the receiver to choose the data rate. However, getting this feedback will take time, and the channel state information may to some extent be outdated before the transmitter can make use of it.

An alternative approach to obtain adaptive information data rate—*without* having channel state information available at the transmitter—is to apply what is known as rateless codes or fountain codes. Such codes can be used to generate an arbitrary amount of redundancy. The basic properties of this family of codes are:

- The number of information bits k is arbitrary
- The code is able to stream an arbitrary amount of code bits based on the information bits
- If the decoder correctly receives **any** k' of the code bits, where k' is only slightly larger than k, it can decode all the transmitted information bits.

Note that it does not matter which k' code bits are received, any subset of sufficient size will do. The term fountain codes is due to the following analogy [10]: "A digital fountain has properties similar to a fountain of water: when you fill your cup from the fountain, you do not care what drops of water fall in, but only want that your cup fills enough to quench your thirst. With a digital fountain, a client obtains encoded packets from one or more servers, and once enough packets are obtained, the client can reconstruct the original [message]. Which packets are obtained [and in which order] should not matter."

The fountain codes typically listed in literature are [10, 11]: Tornado codes, LT (Luby Transform) codes, and Raptor (Rapid tornado) codes. Of these, Raptor codes are the best and newest (published in 2004 [12]). Raptor codes can be encoded and decoded with computational complexity increasing only *linearly* with the number of bits k.

Fountain codes were originally designed for erasure channels, and are therefore sometimes also referred to as erasure codes. In an erasure channel, there are no undetected errors but every bit error has been replaced by an "erasure", i.e., a bit that is in error is erased and disregarded. To convert a real-world noisy channel to an erasure channel, one can e.g. apply CRC: If the CRC check fails on a frame, the entire frame is erased. Note that in this setup, a number of frames with separate CRCs would be combined to generate a code word.

It is however also possible to apply LT and Raptor codes to noisy channels without erasures [11]. This has not been investigated further in the present study.

Fountain codes are applicable to type II hybrid ARQ, where they can be used to generate just as much incremental redundancy as is needed to get the information through, and the receiver can send an ACK back when it has got enough to decode the transmitted information. Fountain codes are particularly suitable to multicast applications, where different receivers experience different error patterns but the transmitted redundancy is still equally usable to all. When the transmitter has received ACK from all the intended receivers, it knows that it has sent enough redundancy and can stop transmitting.

Papers proposing to use fountain codes in underwater acoustic communications include [13, 14] who consider multicast applications, [15] who considers an application where AUVs collect data from underwater sensors, and [16, 17] who consider data-centric storage applications.

4.4 Link Layer Improvement Potential in Networks

When designing link layer protocols for a single point-to-point link, one cannot do much more than discussed in Sects. 4.2 and 4.3. But when more than two nodes are considered, there is a large potential for improving link layer protocols by allowing nodes to cooperate in getting information from its source to destination without errors.

We already mentioned one multinode application, namely multicast, in Sect. 4.3.3. In the present section, we will further investigate possibilities to improve link layer performance in networks.

4.4.1 Topologies

In Fig. 4.2 we show some example network topologies as a reference picture for the discussions in this section. The arrows indicate which way information is transmitted from source(s) to sink(s) in the discussed examples, but of course there

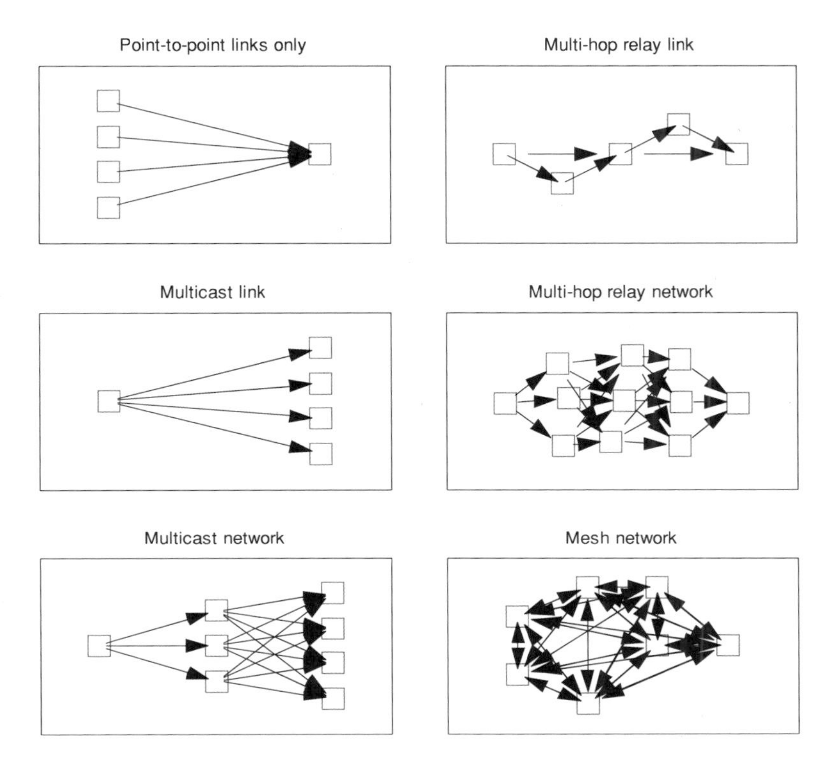

Fig. 4.2 Example network topologies discussed in a link layer context

will be control packets going the other way, and also the same set of nodes may be used to transmit information in other directions.

The upper left panel in Fig. 4.2 is included to illustrate that we may have cases where there are multiple nodes in the network, but hardly any link layer benefit from using several nodes in combination since no information resides in more than one node. Here, there are multiple sources transmitting information to one sink without any relaying.

In a multicast link (mid left panel), a source transmits the same information to several sinks. A multicast network (lower left panel) is similar, but also allows other nodes to participate in getting the information to the sink. As discussed above, one possibility to improve performance in multicast applications is to apply fountain codes.

In a multi-hop relay link (upper right panel), nodes along the way from source to sink are used to forward the information. Since each hop is a fraction of the total distance, the communication performance on each hop will be better (higher data rate, more robust, lower transmit power, etc). However, each hop will add overhead and latency, such that there is a trade-off involved in selecting the number of hops to cover the distance. This trade-off is studied in [18]. In the figure, we have also indicated the fact that some frames may be received also by nodes further down the line, e.g., due to varying channel conditions. Performance may be improved if such opportunities are exploited. Note that the case of a direct link in parallel to a two-hop link is sometimes referred to as a relay network.

In a multi-hop relay network (mid right panel), information travels from source to sinks along several different routes, either in parallel or by switching between different routes as propagation conditions change.

In a mesh network (lower right panel), several nodes are within communication range of another (in the limiting case, all nodes), and any node can act as source or sink. In this case, there are many different ways to get information from source to sink.

When there are moving nodes in the system, it is essential that the applied protocols at all layers—including the link layer—are able to adapt to changes in topology.

4.4.2 Implicit Acknowledgment

In a multi-hop relay link or relay network, it is not required that the forwarding nodes send link layer ACK back to the transmitter at the incoming hop. If for example node A sends a frame to node C via node B, it can hear that the frame is forwarded by node B and then implicitly knows that it was correctly received by node B. This technique is called implicit acknowledgment. Obviously, it will decrease the overhead and overall latency, at the cost of slightly increased protocol complexity.

Papers investigating implicit acknowledgment for underwater acoustic communications include [19–21]. Some of these assume in their analysis that (using our example) node A needs to receive without errors the entire frame forwarded by node B in order to get the implicit acknowledgment. This assumption seems overly conservative to us, as it should be sufficient that node A just observes that B starts sending to C, e.g. by demodulating only the header of the frame forwarded by B.

4.4.3 End-to-End Feedback

In a multi-hop relay link or relay network, one may omit the ARQ feedback on each hop and rather send an ACK/NAK back from the sink to the source along the same multi-hop route as the information took. This is called end-to-end feedback. Due to the long total latency in an underwater multi-hop relay link, this feedback will come very late, and we have not found any papers recommending using end-to-end feedback in underwater acoustic communications. It should, however, not be disregarded completely as there may be cases where it is suitable.

4.4.4 Opportunistic Routing

When the channel propagation and noise conditions are varying in time and space, there will be benefits from applying cross-layer alternatives to standard routing methods at the network layer. One alternative is to apply opportunistic routing, which means to make use of whichever link that happens to be error-free for a given transmission. Information can be forwarded along different routes from transmission to transmission, depending on the actual channel conditions. The problem with opportunistic routing is that measures must be taken to ensure that the network is not congested by flooding.

In [21], Zhuang et al. propose a protocol for opportunistic routing in an underwater multihop relay link. Here, opportunistic routing is combined with implicit acknowledgment in a way that avoids flooding of the network: Assume that nodes A–D are on a line, and that A shall send a message to D. The normal routing would be A–B–C–D. But if the transmission from A can also be heard directly by C without errors, C will take the opportunity and forward the message to D (in effect skipping a hop). When node B hears that C forwards the message, it implicitly knows that the message was correctly received and forwarded by C. Then, B will refrain from forwarding the message. The authors of [21] call this combination of opportunistic routing and implicit acknowledgment "bidirectional overhearing".

Implementation of opportunistic routing would be more complicated in other topologies than the multihop relay link, and we have not found any proposals on this in the underwater communications literature.

4.4.5 Network Coding

A relatively new and active research field that may have potential to increase the performance in underwater networks is network coding. Network coding was first introduced by Ahlswede et al. in 2000 [22] for use in wired networks, and has successively been proposed for wireless radio networks [24], cognitive radio networks [24], and for underwater acoustic networks [25].

An introduction to the field of network coding can e.g. be found in [26]. The main idea is that data flows from multiple sources are combined to increase the network throughput, reduce delay and enhance robustness. In this context network coding offers a paradigm shift with respect to the traditional *store* and *forward* approach [27] by implementing a *store, code* and *forward* technique. Network coding requires each node in the network to store the incoming packets in its own buffer and transmit their combinations (instead of simply forwarding them), where such combining is performed over some finite Galois Field for practical implementations. Hence, by mixing multiple source packets into the same coded packet it is possible to obtain increased throughput efficiency as well as scalability and

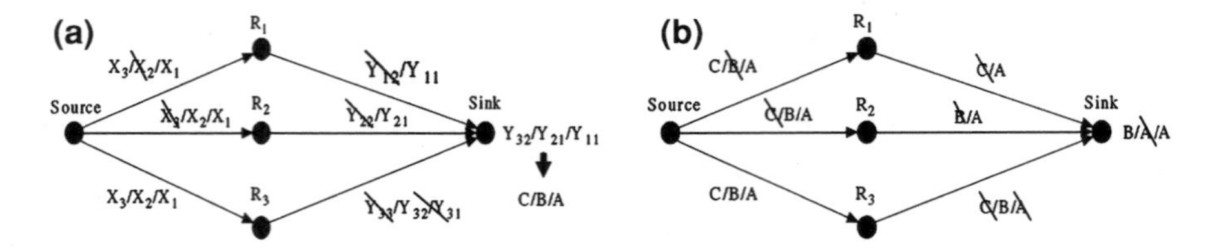

Fig. 4.3 Example of application of network coding, from [34] with permission. **a** Using network coding. **b** Not using network coding

robustness with respect to link failures or bad channel conditions [28–30]. We note that fountain codes (FC, Sect. 4.3.3) use a similar approach, since multiple packets are combined to create a coded output packet (usually this is done by operations in a Galois Field). However, while in the case of FC coding is performed at the link level, over packets generated from the same source node, in the case of network coding packet mixing is performed at the network level, i.e. packets from multiple nodes are combined together. Applications include but are not limited to multicasting [31], broadcasting [29, 30], multiple unicast flows [23, 32], etc.

A typical example to explain network coding is the "butterfly network", which was originally proposed in [22] to show the benefits of network coding. Here, we will briefly describe a slightly different underwater relay network example from [33], which is shown in Fig. 4.3: Three frames A, B, and C are to be transmitted from source to sink. In the right-hand example without network coding, it is assumed that each repeater forwards all successfully received frames towards the sink (i.e. store and forward approach). However, due to frame errors only two of the three frames successfully reach the sink without retransmissions. In the left-hand example nodes are enabled to use network coding, hence relay nodes store and forward different combinations of the successfully received incoming frames, so that the sink can decode all the frames transmitted from the source as long as it receives 3 linearly independent combinations of the source packets, in spite of the number and pattern of channel-induced frame errors (which is the same in both examples).

In network coding, the codes used to combine incoming frames are usually linear codes. For a given topology it is possible to compute optimal coding polynomial coefficients, but when the topology is a priori unknown and/or time-varying it is better to apply random coding coefficients generated in each relay. To this end, [28] was the first contribution to present a practical and distributed solution exploiting random linear network coding. The authors focused on how the coding matrix as well as the information related to the random combination of packets in some finite Galois Field $GF(q)$ can be shared by different nodes at low overhead. This is a crucial aspect for network coding algorithms to work in multi hop radio and acoustic networks. Therefore, random linear network coding results are attractive to underwater environments where robust and distributed communication protocols are vital for the efficient utilization of the available resources. The downside is that additional overhead is required to transmit the coding coefficients, which are not a priori known when random linear network coding is applied. However, in practical settings this overhead is often negligible as shown in [28].

In this context it has been shown that network coding is particularly beneficial in multicast applications. In [34], Chitre and Soh investigate an underwater mesh network, where all nodes act as sources and sinks, and all information from all sources needs to be received by all sinks. They compare a solution based on straightforward ARQ, a solution based on fountain codes, and a solution based on network coding. Their conclusion is that the network coding solution performs better than the fountain code solution, at the cost of computational complexity.

In [35], Lucani et al. investigate the application of network coding to an underwater multihop relay link, where transmissions can be heard and forwarded by more than one downstream node. They combine network coding with implicit acknowledgments, and show the benefits of this combination in the considered topology.

In [36], Chirdchoo et al. take a more general approach to application of network coding to underwater acoustic networks. They investigate in which cases there are benefits to be gained from network coding. In their conclusions, we find the following recommendations: "Network coding is a promising technique that can provide high PDR [packet delivery ratio], high data rate, as well as good energy efficiency. However, network coding can provide such benefits only if it is implemented in a network that satisfies a certain set of criteria. These include: (i) there are enough multiple paths from the source node to the destination node, and (ii) the total number of hops between the source node and the destination node is large enough to allow network coding to be fully utilized."

4.4.6 Collaborative Beamforming and Related Ideas

Collaborative beamforming [37, 38] is a technique suggested for wireless radio ad hoc networks. The idea in its basic form is to transmit the same signal simultaneously and coherently from several relay nodes, which together will form a beam towards the receiver. In this form, the idea does not seem applicable to underwater acoustic networks since the propagation delays will make it hard to synchronize the relays in such a way that the signals reach the receiver simultaneously.

An extension of this idea is to allow a receiver to coherently combine several transmitted signals which are identical but received at different times, from the same transmitter or from several relay nodes. In [39], van Walree and Leus mention the idea of combining several independent transmissions of an identical signal using a multichannel equalizer. They did, however, not test this idea. The idea seems to be applicable to ARQ protocols with several identical retransmissions. In [40], Higley et al. presents an experiment where they coherently combined several transmissions from a towed transducer in a time-reversal fashion. They call this approach "synthetic aperture time-reversal communications".

References

1. Fujiwara T, Kasami T, Kitai A, Lin S (1985) On the undetected error probability for shortened Hamming codes. IEEE Trans Commun 33(6):570–574
2. Kalscheuer JM (2004) A selective automatic repeat request protocol for undersea acoustic links. Master's thesis, Naval Postgraduate School, Monterey CA, USA
3. Rice J, Green D (2008) Underwater acoustic communications and networks for the US Navy's Seaweb program, In: Proceedings of 2nd international conference on sensor technologies and applications (SENSORCOMM), Cap Esterel, France, pp 715–722
4. Rice J, Kalscheuer J (2011) A selective automatic request protocol for through-water acoustic links. In: Proceedings of the 4th UAM 2011, Kos, Greece
5. Stojanovic M (2005) Optimization of a data link protocol for an underwater acoustic channel. In: Proceedings of Oceans Europe 2005, IEEE, vol 1, Brest, France, pp 68–73
6. Tomasi B, Casari P, Badia L, Zorzi M (2010) A study of incremental redundancy hybrid ARQ over Markov channel models derived from experimental data. In: Proceedings of ACM WUWNet, Woods Hole, MA, USA
7. Johnson E, Kenney T, Chamberlain M, Furman W, Koski E, Leiby E, Wadsworth M (1998) US-MIL-STD-188-141B Appendix C—a unified 3rd generation HF messaging protocol. In: Proceedings of HF98, Nordic shortwave conference Fårö, Sweden, pp 5.1.1–5.1.30
8. NATO (2009) Annex C to STANAG 4538 ed. 1: technical specifications to ensure interoperability of an automatic radio control system for HF communication links
9. Chamberlain MW, Furman WN (2003) HF data link protocol enhancements based on STANAG 4538 and STANAG 4539, providing greater than 10 kbps throughput over 3 kHz channels. In: Proceedings of 9th international conference on HF Radio Systems and Techniques Bath, UK, pp 64–68
10. Mitzenmacher M (2004) Digital fountains: a survey and look forward. In: Proceedings of information theory workshop, ITW 2004, IEEE, San Antonio, TX, USA, pp 271–276
11. Bonello N, Chen S, Hanzo L (2011) Low-density parity-check codes and their rateless relatives. IEEE Communications Surveys & Tutorials,13(1):3–26
12. Shokrollahi A (2004) Raptor codes. In: Proceedings of international symposium on information theory, ISIT 2004. IEEE, Chicago, IL, USA, p 36
13. Xie P, Cui J-H (2006) SDRT: a reliable data transport protocol for underwater sensor networks. UCONN CSE technical report UbiNet-TR06-03, University of Connecticut
14. Casari P, Rossi M, Zorzi M (2008) Towards optimal broadcasting policies for HARQ based on fountain codes in underwater networks. In: Proceedings of 5th annual conference on wireless on demand network systems and services, Garmisch-Partenkirchen, Germany, IEEE, pp 11–19
15. Chan CYM, Motani M (2007) An integrated energy efficient data retrieval protocol for underwater delay tolerant networks. In: Proceedings of Oceans 2007 IEEE, Europe, Aberdeen, UK
16. Cao R, Yang L (2010) Decomposed Raptor codes for data-centric storage in underwater acoustic sensor networks. In: Proceedings of MTS/IEEE Oceans 2010, Seattle, WA, USA
17. Cao R, Yang L (2010) Short paper: reliable transport and storage protocol with fountain codes for underwater acoustic sensor networks. In: Proceedings of the fifth ACM international workshop on under water networks (WUWNet), Woods Hole, MA, USA
18. Stojanovic M (2007) Capacity of a relay acoustic channel. In: Proceedings of MTS/IEEE oceans 2007, Vancouver, BC, Canada, IEEE
19. Tan H-P, Seah WKG, Doyle L (2007) A multi-hop ARQ protocol for underwater acoustic networks. In: Proceedings of oceans 2007 Europe, Aberdeen, UK, IEEE
20. Valera A, Lee PWQ, Tan H-P, Liang H, Seah WKG (2009) Implementation and evaluation of multihop ARQ for reliable communications in underwater acoustic networks. In: Proceedings of Oceans 2009 Europe, Bremen, Germany, IEEE

21. Zhuang H, Tan H-P, Valera A, Bai Z (2010) Opportunistic ARQ with bidirectional overhearing for reliable multihop underwater networking. In: Proceedings of Oceans 2010 Asia, IEEE
22. Ahlswede R, Cai N, Robert Li S-Y, Yeung RW (2000) Network information flow. IEEE Trans Inf Theory 46(4):1204–1216
23. Katti S, Rahul H, Hu W, Katabi D, Médard M, Crowcroft J (2006) Xors in the air: practical wireless network coding. ACM SIGCOMM Comput Commun Rev 36(4):243–254
24. Asterjadhi A, Baldo N, Zorzi M (2009) A distributed network coded control channel for multi-hop cognitive radio networks. IEEE Netw 23(4):26
25. Guo Z, Xie P, Cui J-H, Wang B (2006) On applying network coding to underwater sensor networks. In: Proceedings of ACM WUWNet, Los Angeles, CA, USA, pp 109–112
26. Fragouli C, Yves Le Boudec J, Widmer J (2006) Network coding: an instant primer. ACM SIGCOMM Comput Commun Rev 36(1):63–68
27. Eugster P, Guerraoui R, Kermarrec AM, Massoulie L (2004) Epidemic information dissemination in distributed systems. Computer 37(5):60–67
28. Chou PA, Wu Y, Jain K (2003) Practical network coding. In: 41st Allerton conference on communication control and computing, Allerton, IL, USA
29. Fragouli C, Widmer J, Boudec JYL (2008) Efficient broadcasting using network coding. IEEE/ACM Trans Netw 16(2):450–463
30. Asterjadhi A, Fasolo E, Widmer J, Rossi M, Zorzi M (2010) Toward network coding-based protocols for data broadcasting in ad hoc wireless networks. IEEE Trans Wireless Commun 9(2):662–673
31. Ho T, Medard M, Koetter R, Karger DR, Effros M, Shi J, Leong B (2006) A random linear network coding approach to multicast. IEEE Trans Inf Theory 52(10):4413–4430
32. Omiwade S, Zheng R, Hua C (2008) Practical localized network coding in wireless mesh networks. In: Proceedings of IEEE SECON, San Francisco, CA USA, pp 332–340
33. Guo Z, Wang B, Xie P, Zeng W, Cui J-H (2009) Efficient error recovery with network coding in underwater sensor networks. Ad Hoc Netw 7:791–802
34. Chitre M, Soh W-S (2010) Network coding to combat packet loss in underwater networks. In: Proceedings of fifth ACM international workshop on underwater networks (WUWNet), Woods Hole, MA, USA
35. Lucani DE, Médard M, Stojanovic M (2007) Network coding schemes for underwater networks: the benefits of implicit acknowledgement. In: Proceedings of WUWNet 07, ACM, Montreal, QC, Canada, pp 25–32
36. Chirdchoo N, Chitre M, Soh W-S (2010) A study on network coding in underwater networks. In: Proceedings of MTS/IEEE Oceans 2010, Seattle, WA, USA, IEEE
37. Ochiai H, Mitran P, Poor HV, Tarokh V (2005) Collaborative beamforming for distributed wireless ad hoc sensor networks. IEEE Trans Signal Process 53(11):4110–4124
38. Mudumbai R, Barriac G, Madhow U (2007) On the feasibility of distributed beamforming in wireless networks. IEEE Trans Wireless Commun 6(5):1754–1763
39. Paul A, van Walree Geert Leus L (2009) Robust underwater telemetry with adaptive turbo multiband equalization. IEEE J Ocean Eng 34(4):645–655
40. Higley WJ, Roux P, Kuperman WA, Hodgkiss WS, Song HC, Akal T, Stevenson M (2005) Synthetic aperture time-reversal communications in shallow water: experimental demonstration at sea. J Acoust Soc Am 118(4):2365–2372

Chapter 5
Routing

Routing is a fundamental network primitive in any wireless network. Given typical transmit power constraints, it is very unlikely that all nodes in a network are within the transmit range of one another. For this reason, many messages may have to be relayed towards their destination through multiple hops. Other than the clear advantages this strategy brings about in terms of connectivity among far nodes, multihop routing generates two types of overhead: on one hand the messages get replicated throughout the network, as multiple nodes relay the original transmission; on the other hand, the decisions about which node should be a relay require some sort of signalling before routing actually takes place.

Underwater networks are not different from other kinds of wireless networks in this regard. For many scenarios, including harbour patrol, coastline environmental monitoring, etc., the area of operations of the network may span several square kilometers, making single-hop networking infeasible. In addition, the specific features of underwater propagation make multihop topologies more convenient: it has been shown [1] that absorption losses (due to the resonance of pressure waves with salt particles in water) cause a significant attenuation phenomenon, whose entity grows exponentially with distance. This comes in addition to the usual spreading loss factor, which instead depends on distance according to a power law and is found in terrestrial radio transmissions as well. Such attenuation requires that a very high power is used at the transmitter side, in order to cover a long-distance hop while achieving a sufficient Signal-to-Noise Ratio (SNR), and thereby correct reception.

As opposed to what was said above, it was observed [2, 3] that multihop transmissions would yield two different kinds of benefit: they substantially reduce the power required to cover one hop, and at the same time they allow the nodes to perform transmissions on a larger, high-frequency bandwidth, which usually also means higher available bit rates. This may be leveraged in different ways, e.g., by explicitly adapting the bandwidth and power used over a single hop in a way that optimizes energy consumption over a whole end-to-end path. Also this total

R. Otnes et al., *Underwater Acoustic Networking Techniques*,
SpringerBriefs in Electrical and Computer Engineering,
DOI: 10.1007/978-3-642-25224-2_5, © The Author(s) 2012

energy consumption may in some cases be reduced, especially if the total distance is quite large [3].

While the advantages of multihop in terms of power and bandwidth are apparent, the drawbacks of this approach are mainly two: on one hand, more nodes have to take part in packet forwarding, which replicates the same packets multiple times as relaying is in process; on the other hand, a mechanism must be devised to choose relays and identify the routes to walk, which in general requires overhead and messages to be passed (typically back and forth) between the source and the final destination, or between each hop along the path and its neighbours.

In any event, multihop routing faces many challenges in underwater networks: for example, the physics of propagation, in certain environments, may originate links with quickly time-varying performance: in the presence of such rapid variations, it would be difficult to make decisions on which node should be a relay and which should not. In addition, some links may exist only in one direction (e.g., a node close to shore may be able to communicate with a node off shore, but the channel may be too harsh or experience a very high bit error rate in the opposite direction due to the upslope bathymetric profile, which leads to a higher number of reflections of equivalent power). This issue may create substantial lack of balance in routing paths and would need to be addressed.

In the following, we will proceed by a description of general approaches well known in the terrestrial radio environment, before proceeding toward a summary of existing routing protocols and approaches for underwater networks.

5.1 Overview of Routing Protocol Classes

5.1.1 Proactive and Reactive Routing

The distinction between protocols of reactive and proactive nature is among the first to have arisen in the field of ad hoc networks. The details vary among different protocols, but reactive ones, such as AODV [4] and DSR [5], find a route only when there is data to be transmitted, resulting in low control traffic and routing overhead. Proactive protocols like OLSR [6] and DSDV [7] on the other hand, find paths in advance for all source and destination pairs and periodically exchange topology information to maintain them. Many comparative studies can be found in the literature, sometimes offering contrasting results because of the different considered scenarios. However, the general message of these studies is that reactive protocols perform well in mobile networks regardless of the mobility speed, whereas proactive protocols may suffer in high traffic load and high mobility conditions. The main advantage of proactive protocols, instead, lies in the good end-to-end delay achieved, thanks to the fact that the routes are established in advance and continuously maintained. This is confirmed by [8], where the proactive OLSR protocol is shown to perform very well under different mobility and traffic conditions. By way of contrast, reactive protocols perform worse in terms of delay, mainly due to packet discarding while the route is being discovered. As an exception, a notable advantage of reactive protocols based on source routing (such as DSR) is that they actually find more routes as a result of the discovery process, which typically leads to more robustness in case some of the found routes fail.

As a general rule, however, the high amount of messaging required by proactive protocols does not make them very suitable to underwater communications, especially as their overhead signalling may take a large deal of the already small bandwidth available, thereby leaving limited resources for actual data.

Some examples of combined algorithms also exist, that do proactive routing first, and accommodate new nodes or topology changes by allowing nodes to reactively flood route information packets (to within a limited scope).

5.1.2 Geographic Routing

In its simplest instance, geographic routing (or position-based routing) is a mechanism based on progressively advancing a packet toward the destination, i.e., the distance to destination of the node holding the packet should decrease after every successful transmission to the next relay on a multihop path. Geographic routing has many advantages: one among all, it is virtually stateless. In a sufficiently dense network, every node has at least one neighbour eligible to relay a packet according to geographic advancement criteria: this allows very simple greedy routing protocols to be deployed.

Geographic routing has only two major disadvantages: in sparse networks, it may be actually prone to local disconnections in the network, which lead the node currently holding the packet to be the closest to the final destination among its neighbours. In this case, the packet cannot be relayed further according to purely geographic criteria, and the network must cooperate to route it around the local disconnection, until a node is found from which greedy routing can be resumed. Leading a packet out of dead ends actually requires to store some information about the network topology. For example, the nodes on the border of the "connectivity hole" cannot resume greedy routing, and should in general mark themselves as pertaining to the hole boundary; in addition, they should keep track of which node is more convenient to elect as their next hop in order to exit the connectivity hole "optimally" according to some criterion (e.g., minimum number of hops before greedy forwarding can actually be resumed). As the latter kind of information actually represent a network "state" being tracked by a node, topologies with many holes (typically, sparse ones) tend to limit the statelessness of geographic routing.

The second disadvantage of geographic protocols lies in the knowledge of the position of the destination. In static networks this is not an issue, as the destination is located at a well known position, which can even be hardcoded into the firmware of the nodes prior to network deployment. Also, those networks where final destinations are well-known nodes with limited mobility (e.g., gateway buoys floating on the surface) can take advantage of geographic routing. However, in fully mobile networks, or in networks with an a priori unknown topology, the nodes may not be aware of the position of their wanted destination, which can vary over time and thus be different for subsequent packets. Some techniques to make the nodes aware of the network topology by learning from transmission failures can be adopted here [9] but these techniques usually take time, and are thus more suited to static networks. Therefore, either some preliminary location discovery is carried out (e.g., by replicating the route discovery process of DSR, whose messages of broadcast nature also tend to naturally circumvent connectivity holes), or some overlay location service (e.g., [10]) is queried for acquiring the position of the destination. In either case, however, the signalling overhead required may be very relevant, and therefore substantially offset the advantages of geographic routing.

It is worth noting that, for many geographic routing protocols to actually work, the nodes are also supposed to know their own location, which can be obtained by means of localization protocols. In the specific case of underwater networks, the use of acoustic signals may allow better precision than radio signal strength-based ranging in terrestrial radio networks: therefore, some nodes placed at known locations could be used as beacons by other nodes wishing to localize themselves. In other words, these nodes would send signals at regular intervals from which nodes can estimate the received power or time delay from each beacon and infer their own position by means of triangulation or similar techniques. Good candidates for being beacons are surface buoys and bottom-mounted nodes.

A notable technique devised to avoid expensive localization procedures is known as geographic routing without location information [11], and is based on the back-and-forth exchange of messages where, at each received message, each node updates its own coordinates based on some weighted average of the coordinates of its neighbours, before sending a message with its updated location. After a number of cycles, the coordinates can be shown to converge to some stationary topology that makes greedy routing highly likely to succeed. However, the very intensive messaging required before convergence prevents the usage of such techniques in the underwater environment.

5.1.3 Unicast, Broadcast, Multicast, Geocast, Anycast

The terms unicast, broadcast and multicast refer to the scope of a transmission performed by a source node. Unicast usually refers to the fact that only two nodes are involved in the communication. The large majority of the protocols designed for underwater networks operate following a unicast criterion: a specific destination node is targeted by a source, and a message is relayed by the network as having exactly that source and as being directed to exactly that destination. Other nodes may cooperate and relay the message, in general without being interested in the data packets themselves.

By way of contrast, broadcast and multicast have a one-to-many scope. In particular, broadcast is used to pass information to all nodes in a network, and must therefore be designed to deliver packets to many nodes, possibly over multiple hops. A distinction is made between reliable and non-reliable broadcast. Examples of applications for reliable broadcasting include full reprogramming and firmware update on the nodes, or limited update for some operational network parameters; an example of non-reliable broadcasting is a data request via telemetry to any node in the same area (whereby the data reported by any node is equivalently representative, hence it is not required that all nodes receive the request). While the number of communications to be carried out may seem very large (especially in light of the need to recover from possible communication errors), broadcast protocols can make use of the inherent redundancy of the information flowing through the network. For instance, if some nodes lose part of the message being broadcast, some of their neighbours may have completed a correct reception and be ready to relay it further. These nodes would help supply the missing information without need for the other neighbours to explicitly notify that they missed part of the message. The body of work on broadcasting for underwater networks is small if compared to its terrestrial radio counterpart. The only works with relevance to underwater scenarios are [12–16]. In [12–14] the broadcast problem is solved by using specific incremental redundancy codes (named fountain codes, see Sect. 4.3.3) that allow the transmitters to generate infinite amounts of redundancy in case some nodes need help in recovering from errors. Some specific schemes are devised for efficiently using these codes, and their performance in terms of capability to advance toward the boundaries of the network is analyzed. Some practical schemes have also been proposed and studied by means of simulation. In particular, [15] optimizes a push-based broadcast system in order to reduce its response time in the presence of users showing different requirements about the data being broadcast, with specific focus on increasing the efficiency of signalling by reducing the feedback required by the system.

Multicast shares with broadcast the one-to-many nature of the communication, but the scope of the transmission is limited to a subset of the nodes in the network. In general, nodes should explicitly state their interest in the contents being delivered, but it is also possible that the set of receivers is known a priori. For example, considering two cooperating swarms of AUVs performing a patrol mission, the consolidated information gathered by one swarm may be passed on to the other swarm for better coordination during the mission. In this case, the nodes may want this information to reach all nodes in the other swarm. A specific type of multicast where only the nodes located within a certain network area are addressed is called geocasting. This primitive is particularly useful when the location of the desired receiver is known at least approximately. For example, an underwater monitoring system may originate an alarm signal meant to be received at a very specific location, and thereby route the packets toward that location. Unlike in geographic routing, it is not necessary to know the exact coordinates of the destination area to run geocasting: the nodes in its proximity will most probably have better location information regarding that area than the nodes that originated the packets. As a final remark, some applications may require that a packet is routed toward any node in a group or within a certain area. A typical example regards anycasting an alarm message to any ship (or buoy) in the area where the alarm was originated. In this case it is not critical that the message reaches all ships as it is sufficient that it initially reaches any of them and be further relayed from there if required. Similarly, in the AUV scenario mentioned before, a swarm may want to communicate just with any node in the other swarm: the first to get the message will then propagate it to the swarm it belongs to, if required by the application.

5.1.4 Hierarchical Versus Flat Routing

Routing algorithms can further be classified in terms of the presence or absence of a network hierarchy: hierarchical protocols typically establish a relationship of subordination between some nodes and a selected set of master (or head) nodes, whereas in non-hierarchical (or flat) protocols all nodes can take the same roles and are equally functional both from a network organization and a data transport point of view. Hierarchical protocols usually work in two ways: (i) by organizing nodes into clusters or (ii) by setting up a tree spanning the entire network according to some pre-specified criterion. In case (i), the nodes apply some algorithm to figure out whether they should be clusterheads or children nodes. After that, all children nodes will transmit only to the closest clusterhead, whereas clusterheads will set up an efficient way to communicate with one another. In fact, often (i) requires (ii): especially in data collection applications, clusterheads may set up a tree in order to perform such collection more efficiently; the same topology may actually work as a starting point for more advanced data processing algorithms, involving, e.g., data fusion and aggregation. Finally, since clusterheads are usually stressed more than other nodes by collection of data and network organization tasks, an algorithm is applied to rotate the role of clusterhead among the nodes within the same cluster.

Flat algorithms work the opposite way: all nodes are peer, and take equivalent roles in the process of data forwarding whenever required to do so. For example, all greedy geographic algorithms are also flat: every node will make a decision as to which neighbour should be a relay for the current packet. The relay is chosen based on a specific metric which may take into account only advancement toward the destination of the packet, or a combination of advancement and other parameters (e.g., queue of the relay, success rate of the link, size of the neighbourhood of the relay, etc). By way of contrast, connectivity hole avoidance algorithms in geographic protocols are typically not flat: they cause the nodes on the border of the hole (and sometimes their neighbours as well) to store information about the presence of the hole itself, so as to allow incoming packets to be routed along the border of the hole and away from it until greedy geographic routing can be resumed. Hole detection and avoidance algorithms usually require a large amount of signalling overhead, and are thus not very suitable to underwater networks. Notable exceptions include ALBA-R [9] which automatically learns the shape of the holes by leveraging on packet errors, and PAGER [17], a height-based routing algorithm, where nodes artificially increase their distance from the sink upon the occurrence of local minima, so that their neighbours do not recur to them for obtaining advancement (and thus packets do not get stuck at such nodes any more).

5.1.5 Routing in Delay-Tolerant Networks

Delay- and Disruption-Tolerant Networks, commonly known as DTNs, are characterized by their lack of connectivity due to sparseness and mobility of nodes, very long and variable propagation delay, and high bit-error rate of the channel, resulting in a lack of instantaneous end-to-end paths from source to destination. In these challenging environments, popular ad hoc routing protocols such as AODV [4] and DSR [5] fail to establish stable routes. This is because these protocols first try to establish a complete route between the source and the destination and forward data packets only thereafter. However, when instantaneous end-to-end paths do not exist, the routing protocols must adhere to a "store-carry-and-forward" approach, which exploits the mobility of nodes to route data. In this approach, data is moved from the source to the next available node and stored, waiting for an opportunity to forward the data. Mobile nodes can carry the stored data around, and look for opportunities to forward the data to other nodes while moving towards the destination. In this way, packets are gradually moved or carried by nodes, stored and forwarded throughout the network and eventually reach their destination with a certain probability [18–20]. A common technique used in the literature to maximize the delivery probability of the packets is to replicate them to different nodes, so that at least one of the copies will successfully reach its destination with high probability [21].

There are a number of factors to consider regarding routing in a DTN. One of them is whether or not any information about future contacts is available. For example, in interplanetary communications (one of the first scenarios devised for DTN protocols), a planet or moon is the cause of contact disruption, and large distance is the cause of communication delay. However, it is possible to predict the occurrence of future contacts in terms of the time when the communication links will be available, and how long they will last. These types of contacts are known as scheduled or predictable contacts [22]. On the contrary, in disaster recovery networks, there may be no a priori information available about the locations of communicating devices and the duration of the contacts. These types of contacts are defined as intermittent or opportunistic.

Another important consideration is whether or not the mobility of nodes can be exploited for routing, and how to best do so. There are three major cases which classify the level of node mobility in the network.

First, it is possible that no mobile node exists. In this case, contacts appear and disappear solely based on the quality of the communication channel between them as well as on the movement of the obstructing objects.

Second, not all nodes are mobile. Typically, mobiles are referred to as Data Mules [23, 24] in these scenarios, and may be exploited for collecting, storing, carrying and forwarding the data from other static nodes to the destination. An important aspect regarding routing with Data Mules is how to properly distribute data among them since they are the primary source of communication between any two non-neighbouring nodes in the network.

Third, it is possible that the majority of the nodes in the network, if not all, are mobile. In this case, a routing protocol has more options available during contact opportunities, and can choose to utilize any of the strategies as in [18, 25–27]. One example of this type of network is a disaster recovery network where all nodes (generally people and vehicles) are mobile [28]. Another example is a vehicular network where cars, trucks, and buses communicate with one another while moving [18].

Another consideration, recently investigated in the DTN research community, is the availability of network resources. Many nodes, such as mobile phones, are limited in terms of storage, transmission rate, and battery life. Others, such as vehicles on the road, may be much less limited. Routing protocols can make use of this information to determine how data should be transmitted and stored, so as to not overuse the limited channel resources.

Given the typical DTN nature of many underwater networks (especially sparse networks of mobile nodes patrolling a very large area, or sparse networks of sensors which forward their data only to the mobile nodes which happen to pass by), we believe the DTN paradigm to be of particular importance: therefore DTN routing protocols will be devoted a specific section in the following.

One of the most typical ways to create a taxonomy of DTN routing protocols, is based on whether or not the protocol creates message replicas. Routing protocols that do not replicate a message are called forwarding-based, whereas the protocols that replicate messages are called replication-based. This simple, but popular, taxonomy was adopted recently to classify a large number of DTN routing protocols [25].

There are pros and cons to both approaches, and which approach it is preferable to use depends on the application scenario. Forwarding-based routing is generally much less wasteful of network resources, as only a single copy of a message exists in the network at any given time [22, 29]. Therefore, when the destination receives that single copy of the message, no other node has a copy. This eliminates the need for the destination to provide feedback to the network (except for an acknowledgement for the sender) to indicate that other copies of the message can be discarded, as is the case in replication-based approaches. However, forwarding-based approaches usually do not provide sufficiently high message delivery ratios in many DTN scenarios [26]. On the contrary, replication-based routing protocols achieve higher message delivery ratios [18], since multiple copies of the messages exist in the network, and only one (or a few, in some cases as with erasure coding) must reach the destination. Therefore, a typical trade-off here is found between the two approaches, whereby the former spends few resources but may provide only a low probability of correct delivery, whereas the latter tends to spend more resources but also provides better delivery ratios [27]. Moreover, many of the flooding-based protocols are inherently not scalable. A class of replication-based schemes, such as Spray and Wait [26], attempt to find a good working point on this trade-off by limiting the number of replicas of a given message.

It is important to note that the majority of DTN routing protocols are heuristic-based, and non-optimal. This is due to the fact that finding optimal DTN routing procedures and paths is, in general, an NP-hard problem [25].

5.2 Overview of the Most Significant Underwater Routing Approaches

We proceed now by describing some important underwater routing approaches found in the literature. Following the same structure of the introduction, we will deal with non-DTN protocols before moving onward to DTN approaches. We start from [30], where the authors propose a proactive routing protocol performing autonomous network topology establishment, network resource allocation, and traffic flow control. The protocol does so by relying on a central network manager running on a surface node. In order to establish efficient data delivery paths, the network manager (master node) discovers the topology of the network by transmitting probe signals to its neighbours (see Fig. 5.1). Upon the reception of the first probe, the neighbours append their own id to it and flood the probe to their own neighbours. This ultimately builds a connected tree rooted on the network manager. When the probe reaches the border nodes, after a time-out period, they answer back with a topology completion notice to the master node; this notice follows the same path as the topology discovery probes. This way, when the completion notice arrives, the master node will have all the information required to manage multiple traffic sessions across the network. We note that this protocol relies on a centralized controller, and that it is susceptible to single-point-of-failure issues. Moreover, given the significant signalling exchanges for topology discovery, it may not be suitable to mobile and large-scale underwater networks. A possible application, albeit to be verified, is the creation of data collection routing trees in a fixed network (e.g., made of both moored and bottom-mounted sensors), whose only purpose is data sensing and forwarding toward the tree root.

A similar breadth-first, centralized, and proactive routing protocol is used in Seaweb. This "network discovery process" is described in detail in [31]. As in Fig. 5.1, the discovery process is initiated by a master node, and finds routes between the master node and each other node. The cost function minimized by the algorithm seeks link distances as close as possible to a given preferred hop range [31]. The discovery process uses inter-node ranges measured from the propagation

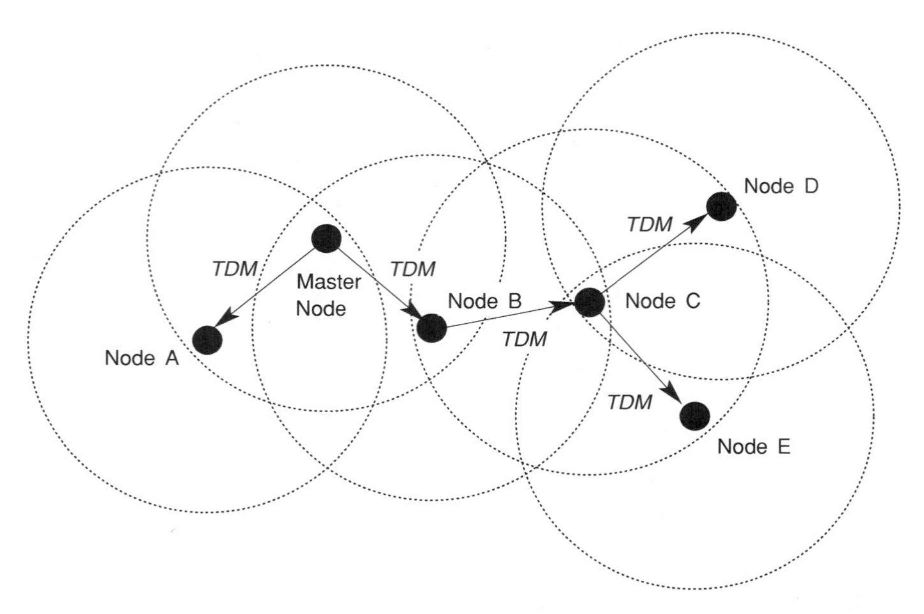

Fig. 5.1 Propagation of topology discovery messages (TDMs) from the master node to border nodes (after [30])

delay between nodes, and the preferred hop range is an input parameter to the process.

In [32], a similar network discovery protocol is described. The protocol is designed for stationary nodes. The mapping is performed by flooding messages to the network, so that the master node eventually obtains the information required to generate the routing tables. An algorithm like this is by nature non-deterministic. There is a chance that a link or node will not be detected due to collisions. With suitable configuration settings and automatic collision avoidance rules, this risk can be reduced to an acceptable level.

In [33], the authors propose two routing algorithms for delay-sensitive and delay-insensitive applications in a 3D underwater environment. The delay-sensitive algorithm is based on virtual circuit routing techniques where multi-hop connections (paths) are established a priori between each source and sink. These multi-hop node-disjoint data paths (primary and backup) are calculated by a centralized controller to achieve optimal performance at the network layer such as minimum delay, energy efficiency and protection against both node and link failures. The delay-insensitive algorithm, conversely, is a distributed geographic solution which aims at minimizing the energy consumption of the communications. In order to do so, nodes forward trains of packets without releasing the channel every time they have a chance to transmit; a cumulative ACK packet is sent for each train to acknowledge correct reception. The algorithm selects the next hop based on an energy metric: in particular those nodes that provide low packet

Fig. 5.2 Vector-based forwarding scheme (after [33])

error rate to the current transfer and which are closer to the destination are preferred.

VBF (Vector Based Forwarding) [34] is a geographic routing protocol where nodes make their forwarding decisions based on the local information available to each node. With reference to Fig. 5.2, the protocol works as follows. Each packet carries the positions of the sender S1, the destination S0, and the relay node which transmits this packet (forwarder). Upon reception of the packet the node computes its relative position by measuring the distance from the forwarder and the AoA (angle of arrival) of the acoustic signal. The node continues forwarding the received packet if it is close enough to the routing vector S1S0 (i.e., located inside a "routing pipe" of given radius W around S1S0). Otherwise it simply discards the packet.

In order to further reduce the number of forwarded packets, and consequently reduce the energy consumption, the authors introduce a "desirableness factor" which enables only a subset of forwarders among the available ones. In more detail, when a node receives a packet, it first determines whether it is close enough to the routing vector. If this is the case, it holds the packet for a time period related to this factor. The idea is to allow the most desirable nodes to forward the packet within shorter time, thereby having other nodes suppress the transmission. This significantly reduces the number of duplicate packets. The evaluation of the protocol shows that it scales well with the number of nodes, and that it is robust and energy efficient. However, it should be noted that this protocol (as all geographic protocols) has an implicit cost, in that the nodes must be aware of their position and of the position of the destination. This requires position estimation techniques, which imply either the presence of dedicated hardware, or the implementation of localization algorithms (e.g., based on beacon nodes that know their own location and transmit such information periodically, so that other nodes can locate themselves via triangulation).

An extension of VBF is the Hop-by-Hop Vector-Based Forwarding protocol (HH-VBF) [35] which aims at overcoming two problems encountered with the native protocol, due to the utilization of a unique source-to-sink vector. This vector allows the creation of only a single virtual pipe between the source and the destination, which may significantly affect the routing efficiency in those network regions where the density of the nodes is significantly low. Moreover, VBF is also very sensitive to the radius of the routing pipe, making it not adequate in a practical network deployment. By allowing nodes to create virtual pipes in a hop by hop fashion it is possible to significantly reduce the impact of these drawbacks, increasing the packet delivery ratio in sparse network deployments while consuming less energy.

A location-aware routing protocol (Focused Beam Routing, FBR) is also proposed in [36]. FBR aims at minimizing the per bit energy consumption via an efficient use of power control. Such protocol is suitable for both static and mobile networks where the source node must be aware of its own location and of that of the final destination.

To show the basics of FBR we refer to Fig. 5.3, where node A is the source and node B the destination. In order for them to communicate, node A sends a request to send (RTS) to its neighbours which contains its location and that of the destination. Such requests are sent using the lowest transmit power, and by increasing it only if necessary (i.e., if the previous transmission did not reach any neighbour). All neighbours of A which receive the RTS first calculate their location relative to the line joining A and B, in order to determine whether they can be candidates for packet relaying. The candidate nodes for FBR are those lying within a cone, as illustrated in Fig. 5.3. A candidate node will then respond to the request with a Clear to Send (CTS). Successively, node A forwards the data packet to this candidate node which repeats the same forwarding procedure, ultimately leading to a multihop path to the destination node B. Simulation results show that a proper coupling of the MAC and routing functionalities with power control provides on-demand route establishment with minimum impact on the network performance.

In [3] the authors focus on the design of energy-efficient routing algorithms for underwater acoustic networks. In this work particular attention has been given to the characteristics of the underwater acoustic signal propagation and the acoustic modem energy consumption profile. The authors make a preliminary analysis quantifying the impact of these characteristics on energy consumption and path delay on simple network topologies. The results of the analysis show that there exists an optimal hop length (the distance between relays) for a given set of the acoustic modem parameters (such as target SNR value and transmit power levels)

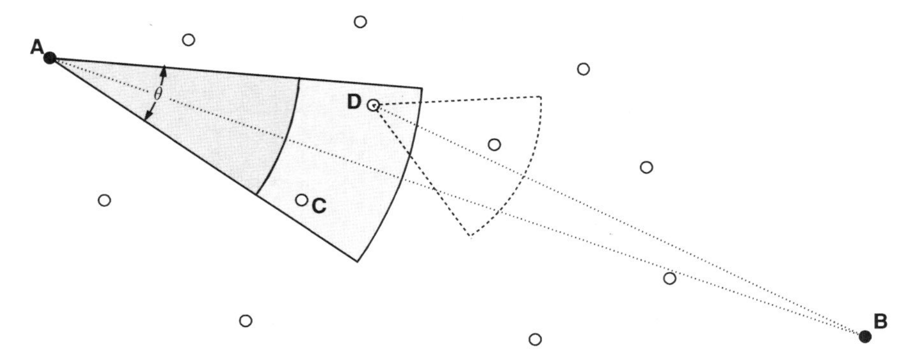

Fig. 5.3 llustration of focused beam routing (FBR): nodes inside the transmitter's cone are candidate relays (after [35])

and a given scenario. Based on the insight provided by the analysis, a geographic routing scheme has been proposed. Such a scheme requires that nodes know their own location and that of the destination, in order to choose the next hop toward the destination. Moreover, nodes are assumed to know the optimal hop distance which assures minimum energy expenditure. Such metric is calculated off-line during network deployment based on the expected features of the application scenario and is assumed to be available to all nodes. The next relay node is then picked among the nodes which guarantee advancement toward the destination and are located nearby the optimal per-hop distance coverage zone. The way nodes are picked is defined by different algorithms, whose performance is compared, among other things, to a centralized optimum algorithm. Simulations results show that using the optimal per-hop distance to pick the relay node assures close to optimum performance in terms of energy efficiency and provides a very good trade-off with respect to throughput and delay in practical scenarios.

While the previous routing solutions require underwater nodes to have an estimate of their own location in order to perform routing, depth-based routing (DBR) [37] makes decisions based only on the nodes' depth, which is especially useful when there is a distinct bottom depth gradient. Hence, it avoids the usage of complex localization techniques and is robust in mobile underwater networks, as depth-related information is easily extrapolated from on-board pressure sensors. DBR is a greedy algorithm, where each node forwards a packet based on its own depth and the depth of the previous sender. More precisely, upon packet reception, the node retrieves the depth of the previous relay (i.e., the depth of the last node to forward the packet, which can be easily stored in the packet itself). In case the node is closer to the surface than the previous hop, it qualifies itself as a candidate for packet forwarding. In dense underwater networks it is most likely that multiple nodes are qualified to forward the packet to the next hop. Hence, if all these nodes do forward the packet, there will be a high risk of packet collision and waste in energy, or otherwise of duplicate packet transmission. In order to reduce the number of candidate nodes that forward the packet, DBR uses Redundant Packet

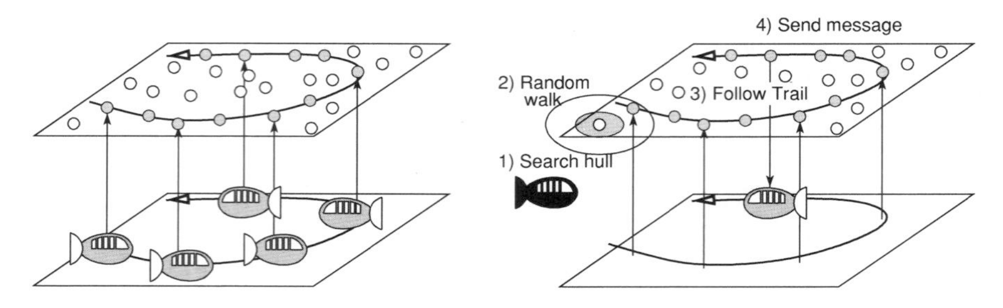

Fig. 5.4 Upper Convex Hull maintenance and routing to the mobile sink (after [38])

Suppression and Timeout parameters to select the best next hop forwarder among all qualified candidates.

In [38] the authors propose a bio-inspired routing protocol (Phero-Trail) for a SEA swarm (Sensor Equipped Aquatic swarm). A SEA swarm operates as a group of sensors which moves with water current and enables 4D monitoring of local underwater activities and aims at reporting critical data in real time to a distant data collection center using such multi hop routing. The proposed routing protocol uses the 2D upper hull of a SEA swarm to store routing information to reach a mobile sink. With reference to Fig. 5.4, such routing information is obtained by having the mobile sink forward its current location to its projection in the 2D upper hull (such sequential location updates are equivalent to a "pheromone trail"). Note that, in this case, sensor nodes do not need to know their own location in order to perform routing, but only their depth, which is used to determine which nodes will become part of the hull servers set. The key idea of Phero-Trail, i.e., to store the mobile sink's location updates in the upper hull, allows sensor nodes to find the pheromone trail every time they are required to forward a packet to the mobile sink. To do so they simply forward the packet vertically upwards to the node on the projected position of the convex hull plane where, after following a random walk over this plane, it finds the pheromone trail which will finally forward the packet to the mobile sink.

Following the same trend as [37, 38], pressure-based routing is proposed in [39], where packet forwarding decisions are also made locally based on measured pressure levels (which depend on depth). Vertical transmissions are preferred in this case. However, in order to avoid situations where a node finds no neighbours with a lower pressure level (a problem similar to the dead-end issue in geographic forwarding protocols) the authors propose a recovery method which guarantees an escape route every time the forwarding node encounters a void region in the swarm (nodes LM1 and LM2 in Fig 5.5). The discovery of such recovery route is performed using 2D flooding over the void floor surface instead of using simple 3D flooding which would require additional overhead for route establishment. As for the selection of the forwarding nodes, the decision is based on the prioritization of the forwarding candidates. Such prioritization is based on a metric which approximates EPA (Expected Packet Advance) which is the normalized sum of the

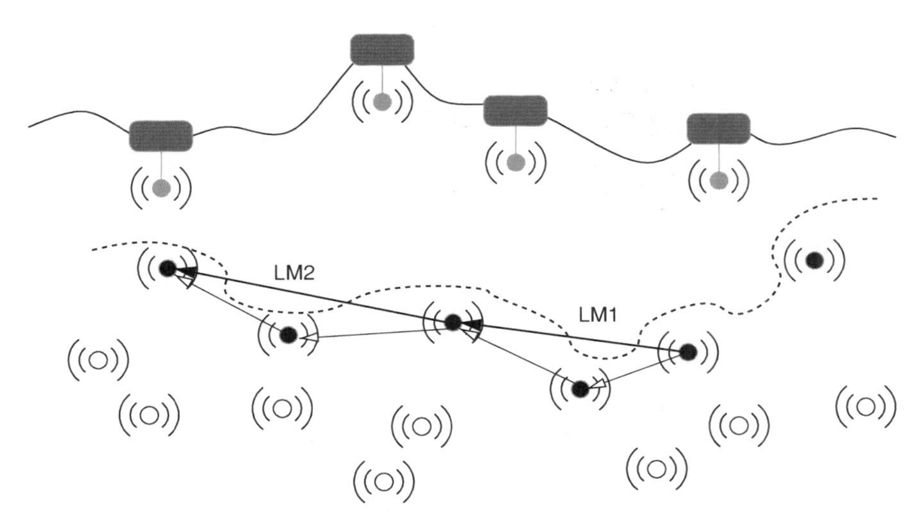

Fig. 5.5 Sea Swarm architecture (after [39])

advancements allowed for by neighbouring nodes. In this case, the advancement of a neighbour takes into account both the distance from the destination and the probability of packet delivery. In other words, all neighbours receiving a packet will access with a certain priority, given by how close they are to the destination (the closer to the destination the higher the priority) and how good the channel quality is (higher packet delivery probability). A node will forward the packet when all nodes with higher progress to destination (higher priority) have failed to do so. This is achieved by setting a back-off timer proportional to the destination distance which allows a node to overhear eventual packets (data or ACKs) sent by higher priority nodes, hence suppressing their transmission.

5.3 Overview of DTN Routing Protocols and Approaches

We proceed by describing routing approaches which have been specifically designed for DTNs. Vahdat and Becker present a routing protocol for intermittently connected networks called Epidemic Routing [21]. Epidemic routing is flooding-based in nature, as nodes continuously replicate messages to the newly discovered contacts that do not already possess a copy of the message. In the simplest case, epidemic routing coincides with flooding; however, more sophisticated techniques may limit the number of message replicas. This protocol relies on the theory of epidemic algorithms by doing random pair-wise exchange of messages among nodes as they get in contact with each other to eventually deliver messages to their destinations. Hosts always buffer messages, even if there is currently no path available to the destination. An index of these messages, called summary vector is maintained by the nodes, and when two nodes meet they first exchange summary vectors. After that, each node can determine if the other node owns any previously unseen messages. In that case, the node requests these messages. Messages must also contain a hop count field, which determines the maximum number of hops a message can travel; this field can also be employed to limit usage of network resources. Messages with a hop count of one will only be delivered to their final destinations.

The resource usage of this scheme is regulated by the hop count set in the messages, and the available buffer space at the nodes. If these are sufficiently large, the message will eventually propagate throughout the entire network. Vahdat and Becker showed that by choosing an appropriate maximum hop count, the delivery ratios can be maintained at a high value, while the resource utilization is decreased, at least in the scenarios used in [21].

Epidemic routing is basically resource-exhaustive because it deliberately makes no attempt to eliminate message replications so as to improve the delivery probability of messages. However, in realistic situations, encounters are not totally random. The Probabilistic Routing Protocol using History of Encounters and Transitivity (PROPHET) [40] attempts to exploit the non-randomness of real-world encounters by maintaining a set of probabilities for successful delivery to known destinations in the DTN, and replicating messages during opportunistic encounters only if the node that does not possess the message appears to have a better chance of delivering it to the destination. Results show that PROPHET clearly outperforms epidemic routing in scenarios where mobility actually follows specific patterns, which the protocol can track and exploit. Even in the random mobility scenario, the performance of PROPHET in terms of delivery ratio and delay is comparable to that of epidemic routing, but with lower communication overhead. PROPHET has been incorporated into the reference implementation maintained by the IRTF DTN Research Group (http://www.dtnrg.org/) and the current version is documented in an Internet Draft (http://tools.ietf.org/html/draft-irtf-dtnrg-prophet) [41].

MaxProp [18] was developed to address routing in vehicular DTNs. The protocol is flooding-based in nature. The refinements brought about by MaxProp lie in determining which messages should be transferred first, and which should be dropped when the buffer of a node is full. In essence, MaxProp maintains an ordered queue based on the destination of each message, sorted by the estimated likelihood that a future path to that destination is found.

When two nodes meet, they first exchange their estimated node-meeting likelihood vectors. Ideally, each node will have an up-to-date vector from every other node. With these vectors at hand, the node can then compute a shortest-path via a depth-first search where path weights indicate the probability that the link does not occur. These path weights are summed to determine the total path cost, and are computed over all possible paths to the desired destinations. The path with the least total weight is selected for that particular destination, and its cost recorded. The messages are then ordered by destination costs, and transmitted, or otherwise dropped (again in order of cost) in case the buffer is full.

In conjunction with the core routing algorithm, MaxProp allows for several complementary mechanisms, each assisting to improve the message delivery ratio and to reduce delivery delay. First, acknowledgements are injected into the network by the final destinations which successfully receive a message, in order to have nodes delete extra copies of that message from their buffers. Second, messages with low hop-counts are given higher priority: this helps promote initial rapid message replication to provide new messages a "head-start". Third, each message maintains a "hop list" indicating the nodes it has previously visited, so to ensure that it does not revisit the same node.

Spray and Wait [26] is a DTN routing protocol that attempts to gain the delivery ratio benefit of replication-based routing while keeping resource utilization low as in forwarding-based routing. Spray and Wait achieves resource efficiency by setting an upper bound on the number of copies per message allowed in the network. The Spray and Wait protocol is composed of two phases, namely, the spray phase and the wait phase. When a new message is generated, the protocol assigns a number L which is attached to that message indicating the maximum allowable copies of the message throughout the network. During the spray phase, the source of the message "sprays" or transfers one copy to L distinct relays. When a relay receives the copy, it enters the wait phase, whereby it simply stores that particular copy of the message until the final destination of the message is encountered for direct delivery.

There are two main versions of the Spray and Wait routing protocol, respectively known as Vanilla and Binary. These two versions differ in the mechanism employed to "spray" the L copies of a message. The simplest way to achieve this, called Vanilla, is to transmit a single copy of the message to the first L distinct nodes it encounters after the message is generated. The second version, referred to as Binary, works as follows: the source node starts with L copies of the message. It transfers L/2 of the copies to the first node it encounters. Each of the nodes then transfers half of the copies they have to future nodes they meet that have no copy of the message. When a node eventually gives away all of its copies, except for one, it switches to the wait phase where it waits for a direct transmission opportunity with the final destination of the message. The advantage of the Binary version is that messages are disseminated much faster than the Vanilla version. In fact, the authors analytically proved that Binary Spray and Wait is optimal in terms of minimum expected delay among all schemes of Spray and Wait, assuming that node movement is a process with independent and identically distributed time samples.

RAPID [25] stands for Resource Allocation Protocol for Intentional DTN. The authors of RAPID argue that prior DTN routing algorithms affect performance metrics, such as average delay and message delivery ratio, only in an incidental fashion. The goal of RAPID is to intentionally optimize a single routing metric among average delay, number of missed reception deadlines, and maximum delay.

The core of the RAPID protocol is based on the concept of utility function, which assigns a utility value, depending on the routing metric to be optimized. The utility is defined as the expected contribution of the packet to the routing metric.

RAPID replicates first those packets that result in the locally highest utility function increase. RAPID, like MaxProp, is also flooding-based, and will therefore attempt to replicate all packets if network resources such as storage and bandwidth so allow. The overall protocol comprises the following steps: (i) metadata is exchanged to help estimating the packet utilities; (ii) direct delivery, as packets destined to neighbouring final destinations are immediately transmitted; (iii) replication, as packets are replicated based on the marginal utility, that is the change in utility over the size of the packet; (iv) termination, as the protocol ends when contacts break or all packets have been replicated.

Finally, it should be noted that one of the seminal analytical papers in the field of delay-tolerant networking is [42], where the authors introduce a simple stochastic model to evaluate the performance of DTN protocols operating through "store-carry-forward". The proposed model is generic and requires only two input parameters: (1) the number of nodes in the network, and (2) the intensity parameter of an exponential distribution modelling the inter-meeting times between two random mobile nodes. Both a closed-form expression and an asymptotic approximation (as a function of the number of nodes) of the expected message delay (defined as the time needed to transfer a message from its source to its final destination) are derived. As an additional outcome, the probability distribution function is obtained for the number of copies of the message at the time when the message is delivered to its destination. These derivations are done for two routing protocols: (1) the two-hop multi-copy, and (2) the unrestricted multi-copy protocols. In the two-hop multi-copy protocol, the source node copies the message to all nodes it meets along its way. Any node which has received a copy of the message from the source node may only deliver it to the destination node. Whereas, in the unrestricted multi-copy protocol, the source node copies the message to all nodes it meets (as in the two-hop multi-copy protocol), but any node that carries the message may in turn copy it to all nodes encountered along its path.

The results in [42] demonstrate the ability of the proposed analytical model to predict the expected message delay under both the two-hop multi-copy and the unrestricted multi-copy protocols for different mobility patterns, across any number of nodes and communication radii.

As an additional result, the authors argued that for all three mobility models, the inter-meeting time intensity is well-approximated by a linear function of the transmission range. This approximation is valid as long as r is not "too large" with respect to the size of the area in which the nodes move. On the other hand, when the number of nodes increases, the transmission range should decrease to prevent excessive interference.

5.4 Conclusions Regarding Routing

The research activities on routing protocols for underwater networks are comparatively recent, with respect to well-established research on point-to-point communications and signal processing techniques, and even to MAC protocol design. Many approaches are straightforwardly derived from terrestrial radio communications and not directly designed for the underwater channel, with a few notable exceptions. Even DTN protocols, which are probably well suited to sparse mobile underwater networks, have been mostly applied to mobile radio networks, and have rarely been tested with underwater communication parameters.

Therefore, several steps are required in this direction, the most important being a closer matching between underwater propagation and routing policies. For example, in those environments featuring specific phenomena such as downward or upward refraction, surface channels, convergence and shadow zones, etc., it would be interesting to exploit the knowledge about the occurrence of these events, so that routing protocols may autonomously adapt to the channel and improve the communications performance.

References

1. Urick R (1983) Principles of underwater sound. McGraw-Hill, NewYork
2. Porto A, Stojanovic M (2007) Optimizing the transmission range in an underwater acoustic network. In: Proceedigs MTS/IEEE Oceans, IEEE, Vancouver, BC, Canada
3. Zorzi M, Casari P, Baldo N, Harris A (2008) Energy-efficient routing schemes for underwater acoustic networks. IEEE J Selected Areas in Comm 26(9):1754–1766
4. Perkins C, Belding-Royer E, Das S. Ad hoc on-demand distance vector (AODV) routing. http://www.ietf.org/rfc/rfc3561.txt IETF RFC
5. Johnson DB, Maltz AM (1996) Mobile Computing. Kluwer Academic Publishers
6. Clausen T, Jacquet P Optimized link state routing protocol (OLSR) http://tools.ietf.org/html/rfc3626 IETF RFC
7. Perkins CE, Bhagwat P (1994) Highly dynamic destination-sequenced distance-vector routing (DSDV) for mobile computers. ACM SIGCOMM Comput Commun Rev 24(4):234–244
8. Mbarushimana C, Shahrabi A (2007) Comparative study of reactive and proactive routing protocols performance in mobile ad hoc networks. In: Proceedings of AINAW
9. Casari P, Nati M, Petrioli C, Zorzi M (2007) Efficient non-planar routing around dead ends in sparse topologies using random forwarding. In: Proceedings of IEEE ICC, Glasgow, Scotland
10. Ratnasamy S, Karp B, Yin Li, Yu F, Estrin D, Govindan R, Shenker S (2002) GHT: a geographic hash table for data-centric storage. In: Proceedings of ACM WSNA, Atlanta, GA, USA
11. Rao A, Ratnasamy S, Papadimitriou C, Shenker S, Stoica I (2003) Geographic routing without location information. In: Proceedings of ACM MobiCom, San Diego, CA, USA
12. Casari P, Harris AF (2007) III Energy-efficient reliable broadcast in underwater acoustic networks. In: Proceedings of ACM WUWNet, Montréal, QC, Canada
13. Casari P, Rossi M, Zorzi M (2008) Towards optimal broadcasting policies for HARQ based on fountain codes in underwater networks. In: Proceedings of IEEE/IFIP WONS Garmisch-Partenkirchen, Germany, pp 11–19
14. Casari P, Rossi M, Zorzi M (2008) Fountain codes and their application to broadcasting in underwater networks: Performance modeling and relevant tradeoffs. In: Proceedings of ACM WUWNet, San Francisco, CA, USA
15. Nicopolitidis P, Papadimitriou GI, Pomportsis AS (2010) Adaptive data broadcasting in underwater wireless networks. IEEE J Ocean Eng 35(3):623–634

16. Mirza D, Lu F, Schurgers C (2009) Efficient broadcast MAC for underwater networks. In: Procedings of ACM WUWNet, Berkeley, CA, USA
17. Zou L, Lu M, Xiong Z (2005) PAGER: A distributed algorithm for the dead-end problem of location-based routing in sensor networks. IEEE Trans Vehicular Technol 55:1509–1522
18. Burgess J, Gallagher B, Jensen D, Levine BN (2006) MaxProp: routing for vehicle-based disruption-tolerant networks. In: Proceedings of IEEE InfoCom. IEEE
19. Juang P, Oki H, Wang Y, Martonosi M, Peh LS, Rubenstein D (2002) Energy-efficient computing for wildlife tracking: design tradeoffs and early experiences with ZebraNet. In: Proceedings of ASPLOS, San Jose, CA, USA, pp 96–107
20. Chaintreau A, Hui P, Crowcroft J, Diot C, Gass R, Scott J (2007) Impact of human mobility on opportunistic forwarding algorithms. IEEE Trans Mobile Comput 6(6):606–620
21. Vahdat A, Becker D (2000) Epidemic routing for partially connected ad hoc networks. Technical Report CS-2006-06. Department of Computer Science, Duke University
22. Jain S, Fall K, Patra R (2004) Routing in a delay-tolerant network. In: Proceedings of ACM SIGCOMM, Portland, OR, USA, pp 145–148
23. Jea D, Somasundara A, Srivastava M (2005) Multiple controlled mobile elements (data mules) for data collection in sensor networks, chapter in "Distributed computing in sensor systems". Lecture Notes in Computer Science, Springer, vol 3560, pp 244–257
24. Shah RC, Roy S, Jain S, Brunette W (2003) Data mules: Modeling a three-tier architecture for sparse sensor networks. In: Proceedings of IEEE WSNPA, Anchorage, AK, USA, pp 30–41
25. Balasubramanian A, Levine BN, Venkataramani A (2007) DTN routing as a resource allocation problem. In: Proceedings of ACM SIGCOMM, ACM, pp 373–384
26. Spyropoulos T, Psounis K, Raghavendra CS (2005) Spray and wait: an efficient routing scheme for intermittently connected mobile networks. In: Proceedings of ACM SIGCOMM workshop on delay-tolerant networking. pp 252–259
27. Spyropoulos T, Psounis K, Raghavendra CS (2007) Spray and focus: efficient mobility-assisted routing for heterogeneous and correlated mobility. In: Proceedings of IEEE PerCom, White Plains, NY, USA, pp 79–85
28. Nelson SC, Harris AF, Kravets R (2007) Event-driven, role-based mobility in disaster recovery networks. In: Proceedings of ACM CHANTS, Montréal, QC, Canada, pp 27–34
29. Henriksson D, Abdelzaher TF, Ganti R (2007) A caching-based approach to routing in delay-tolerant networks. In: Proceedings of ICCCN, Honolulu, HI, USA, pp 69–74
30. Xie GG, Gibson JH (2001) A network layer protocol for UANs to address propagation delay induced performance limitations. In: Proceedings of MTS/IEEE OCEANS, Honolulu, HI, USA, pp 2087–2094
31. Ong CW (2008) A discovery process for initializing ad hoc underwater acoustic networks. Master's thesis, Naval Postgraduate School, Monterey, CA
32. Rustad H (2009) A lightweight protocol suite for underwater communication. In: Proceedings of 2009 international conference on advanced information networking and applications (workshops), Bradford, UK, pp 1172–1177
33. Pompili D, Melodia T (2005) Three-dimensional routing in underwater acoustic sensor networks. In: Proceedings of ACM PE-WASUN, Montréal, QC, Canada
34. Xie P, Cui JH, Lao L (2005) VBF: vector-based forwarding protocol for underwater sensor networks. In: Proceedings of IFIP Networking, Waterloo, ON, Canada
35. Nicolaou N, See A, Cui JH, Maggiorini D (2007) Improving the robustness of location-based routing for underwater sensor networks. In: Proceedings of MTS/IEEE OCEANS. IEEE
36. Jornet JM, Stojanovic M, Zorzi M (2010) On joint frequency and power allocation in a cross-layer protocol for underwater acoustic networks, IEEE Journal of Oceanic Engineering 35(4):936–947
37. Yan H, Shi Z, Cui JH (2008) DBR: Depth-based routing for underwater sensor networks. In: Proceedings of IFIP Networking'08

38. Vieira FML, Lee U, Gerla M (2008) Phero-trail: A bio-inspired location service for mobile underwater sensor networks. IEEE Journal on Selected Areas in Communications 28(4):553–563
39. Lee U, Wang P, Noh Y, Vieira FML, Gerla M, Cui JH (2010) Pressure routing for underwater sensor networks. In: Proceedings of IEEE InfoCom, San Diego, CA, USA, pp 1676–1684
40. Lindgren A, Doria A, Schéln O (2003) Probabilistic routing in intermittently connected networks. In: Proceedings of ACM MobiCom, San Diego, CA, USA
41. Lindgren A, Doria A, Davies E, Grasic S (2010) Probabilistic routing protocol for intermittently connected networks http://tools.ietf.org/html/draft-irtf-dtnrg-prophet-07 IETF Internet Draft
42. Groenevelt R, Nain P, Koole G (2005) The message delay in mobile ad hoc networks. Perform Eval 62(1–4):218–228